The Eucalyptus

Center Books in Natural History

Shannon Davies
Consulting Editor

George F. Thompson
Series Founder and Director

Published in cooperation with the Center for American Places, Santa Fe, New Mexico, and Harrisonburg, Virginia

The *Eucalyptus*

A Natural and

The Johns Hopkins University Press / Baltimore and London

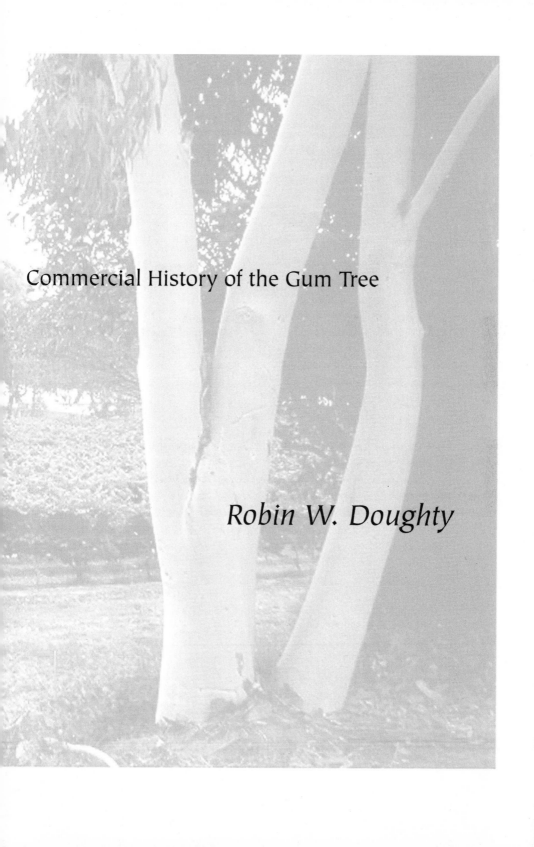

Commercial History of the Gum Tree

Robin W. Doughty

This book was brought to publication with the generous assistance of a University Cooperative Society Subvention Grant awarded by the University of Texas at Austin.

The Johns Hopkins University Press
2715 North Charles Street
Baltimore, Maryland 21218-4363
www.press.jhu.edu

Library of Congress Cataloging-in-Publication Data will be found at the end of this book.
A catalog record for this book is available from the British Library.

ISBN 0-8018-6231-0

Contents

Illustrations appear after pages 11, 59, 87, and 121

Preface and Acknowledgments

Protests about planting eucalyptus trees southeast of Bangkok rocked the Thai government in 1991, as villagers raided nurseries and ripped out seedlings. Local people were objecting to government support for planting eucalyptus on public lands, especially because authorities intended to turn the trees into commercial pulp to feed the expanding international market for paper and newsprint.

Similar militancy took place about the same time in Karnataka, India, after international groups, including the World Bank, collaborated with national and state agencies in covering thousands of hectares with quick-growing eucalyptus plantations. Again, rural folk felt excluded. In Spain and Portugal, farmers and others have also yanked seedlings out of the ground. There, protesters characterized the multiplication of gridlike stands of these hardwoods as "capitalist" or "fascist," thus equating major plantings with right-wing politics during the Franco era.[1]

Initially, this hostility took officials by surprise. Foresters and agronomists pride themselves on resolving problems, not creating them; on opening up opportunities, not closing them down. The reason for recommending eucalypts was the range of economic benefits they supplied in terms of timber, fuel, pulp, and industrial or medicinal oils. Eucalypts produce wood quickly, especially in moist and fertile soils, and many species sprout from the base after being cut. Accordingly, scientists have studied them to learn ways of manipulating and managing the most adaptable and useful species. Researchers introduce, establish, and promote a range of similar plants, many of which are exotics—that is, transplanted into areas where they do not occur naturally. The basic objective is to meet an ever-increasing worldwide demand for fuel, timber, pulp, and other wood-related products.

Since World War II, managers of forestry programs aimed at planting areas in the subtropics and tropics have frequently turned to easy-to-set, fast-growing, resilient tree species like eucalypts, which can adjust to a range of biophysical conditions. Eucalypts can be made to flourish on lands stripped and degraded by mining operations, poor husbandry, or monocultural plantations intended chiefly for export. Some species spring up on areas that have been switched out of agriculture due to prolonged drought or soil erosion. Their proven adaptability and ease of handling have caused a dozen or more species to become an important element in land-use and development programs in the warmer zones of Asia, Africa, and Latin America.

The popularity of eucalyptus trees and the ubiquity of their use has also turned them into targets. Environmentalists and other critics argue that eucalypts typify recent trends in large-scale, modern agro-forestry and tree farming. In their eyes, the trees are components of forestry schemes often drawn up by distant officials, who collaborate with international corporations and entrepreneurs in maximizing cash profits by setting up industrial plantations. In addition to disputing the longer term economic advantages of growing eucalypts, opponents also draw attention to the social costs of these hardwoods and to their controversial impacts upon the environment.

Over the past 150 years, these "gum trees" have spread far beyond the boundaries of their native ranges in Australia and a few neighboring islands, and their rate of spread has increased with time. Louder opposition has been voiced over the last twenty years or so, in part because the plant genus is now so widespread and abundant, and in part because it is grown so intensively. Today, eucalypts sprout in a mosaic of environments in both the Old and New Worlds, from sea level to more than 3,000 m (10,000 ft). Millions of hectares have been given over to them, in the form of plantations, forests, woodlots, fence-rows, and shade trees along roads and city streets. Eucalypts anchor the banks and levees of canals. They ornament estates, parks, and playgrounds, and they serve as windbreaks for field and orchard crops.

Scientists have devoted a great deal of research time and energy to the cultivation and use of eucalypts. The tree has a certain mystique among professionals in tropical and subtropical regions both within Australia and elsewhere. Advances in seed procurement and provenance, and in the vegetative cloning of key species, have enabled foresters to select and breed superior stock that fixes cellulose quickly and efficiently and may be directed to specific outlets. This trend toward intensified wood production is not lost on critics, who argue that research into other woody species, most notably native plants, has been subordinated to the study of high-status trees, such as eucalyptus, which land managers can use as a kind of "off the shelf" remedy both for increasing wood production and for making degraded areas useful again.

It has been said that "no Australian genus has been more thoroughly studied by systematic botanists" than the eucalypts. The genus has become the "tree of choice" for many tree scientists, who have established just how well it adapts to a range of climate and soils beyond its native ground. Almost all the most commonly grown eucalypts germinate readily, have high rates of seedling survival, grow rapidly, coppice well, and yield a range of wood products, such as fuel, charcoal, timber (including posts, poles, sawn lumber, and plywood), wood pulp, medicine and industrial oils, and honey. With some justification, it would seem, proponents are baffled by the drumbeat of criticism and hostility over the expansion of the land area given over to this genus, and they construe objections to its adoption and expansion as unwarranted and ill-founded.[2]

This book sets these issues into historical context. First, it describes the discovery, introduction, and establishment of eucalypts beyond Australia and documents how a cadre of nineteenth-century plant experts and enthusiasts came to express a high regard for the genus and to popularize its adoption. Then, it follows the path taken by industrial, governmental, and international agencies in promoting specific uses for eucalypts. Representatives and employees of these groups quickened the tempo of research, helped establish the plants in various countries on a secure footing, and expanded the geographical range of many species. The book concludes by detailing the controversial and sensational aspects of growing eucalypts outside Australia, specifically the so-called costs to the environment and society of those trees preferred by plantation forestry programs. Regard for the plants has swung from an original mood of curiosity, acceptance, and enthusiasm to an attitude of cynicism, distrust, and hostility, as expressed in recent debates about continued planting. The environmental lobby has pilloried the inclusion of many eucalypt species in schemes for environmental refurbishment and economic development.

In the 1800s and earlier, colonial interests in western Europe began to experiment with a range of new or little-known plants from distant possessions. Initial interest in the plant genus *Eucalyptus* reflected a desire to collect, catalogue, study, and assess both the economic and aesthetic characteristics of new or unusual woody plants. Promotion in the source region, Australia, led to a push into western Europe and North Africa. Seeds also arrived in India and the United States, as botanists, foresters, and others emphasized the potential benefits of growing these plants. Early growers in Europe and the United States welcomed eucalypts. Not only were the novel trees attractive, they also promised to satisfy a range of material needs. Commercial nurserymen and horticulturists, and later public health officials, promoted the popularity and hastened the spread of the genus.

It is instructive to compare early hopes and expectations, freely expressed in journals, pamphlets, newspapers, and the like, with recent opinions, which are guarded and frequently defensive. Initial enthusiasm for eucalypts tapered off after a generation or so as boosters were forced to assess the performance of the trees. They had to recognize that while some eucalypt species had succeeded in certain places, others had not lived up to expectations. Still other species appeared to have outlived their usefulness or outstayed their welcome. Did a fad turn into a fiasco? Or did pragmatism, detailed trials, and prolonged experiments manage to turn heady excitement into a record of sober accomplishment? The answer you will hear depends on whom you ask and when and where you pose the question. This book explores the debate and places in a broader perspective the issue of whether or not to plant eucalypts. By examining the history of introductions, the book compares early information and opinions about what the trees were and what they could do with recent complaints about the accelerated rate of cultivation and spread.

From uncertain and small-scale beginnings at the hands of a few connoisseurs and scientists, eucalyptus trees gained such popularity as to make them in time reportedly the most widely planted hardwood in the world. This reputation is likely to grow as foresters, planners, government officials, and corporate executives support, sponsor, and subsidize the introduction, spread, and use of a selected number of species as a source of valuable wood, which can be grown responsibly as a renewable crop.

Clearly, there are important economic benefits to be realized from growing eucalypts. Growers and promoters have certain expectations in mind that are linked to the overall ecology of eucalypts, and they have specific markets in mind for the wood products that the trees provide. This book explains these expectations and outlets and places them in the context of issues in forest science, sustainable development, and participatory democracy, most notably in the Third World.

*

Many people have offered encouragement and advice about eucalypts. My interest in introductions and uses began in my graduate-student days in Berkeley, California, and grew as I came across Australian trees in various places in Europe, North Africa, Latin America, and Australia. I thank senior forest officer, Food and Agriculture Organization of the United Nations, J. B. Ball and tree specialist David Kleinig for important information. I am most grateful to forestry and geography colleagues at the Australian National University, Canberra, especially Peter Kanowski, who invited me to base myself in the Department of Forestry there. I am indebted to the faculty members, John Banks, Jurgen Bauhus, Cris Brack, Geoff Cary, Philip Evans, Ann Gibson, Irene Guijt, Mike Slee, Mick Tanton, Chris Tidemann, and Brian Turner. Ross Florence was most kind in guiding me in the field. Commonwealth Scientific and Industrial Research Organization (CSIRO) Forestry and Forest Products experts Doug Boland, Steve Midgley, and John Doran, and emeritus professor of botany, Australian National University, Lindsay D. Pryor generously gave me the benefit of their years of experience working with eucalypts in forestry programs in Australia and overseas. Librarians at CSIRO directed me toward data and information.

I am especially indebted to CSIRO Forestry and Forest Products honorary fellow Ken Eldridge, who took time to discuss various issues about planting and who made important comments about the manuscript. I thank Ken Johnston and Bob Wasson in the Department of Geography for their support. Gerard and Marion Ward made helpful suggestions. I am indebted to Brian and Fiona Findlayson and the Geography Department in Melbourne for their hospitality as I explored materials concerning von Mueller. Muelleriana expert Sara Maroske provided valuable insight and guidance about that state's first government botanist. Ian Tyrrell, University of New South Wales, provided information about California. I am grateful to Shirley Kral for her friendship and encouragement during my stay in Australia, and to Peter Milburn for his infectious enthusiasm in the field.

I am grateful to the University of Texas, which has supported this project, most recently with grants from the University Research Institute that assisted research and publication. I thank Richard Meier, Office of Vice-President for Research, for assistance with a University Cooperative Society Subvention Grant awarded by the University of Texas at Austin to help defray publication costs. Provost Sheldon Ekland-Olson graciously allowed me time off as a Dean's Fellow to conduct research in Australia. The Australian Studies

Center, under the aegis of John Higley assisted by Robert Cushing, encouraged this book project and helped with travel funds.

Geography colleagues have been most supportive. Robert Holz lent his expertise about eucalypts in the Middle East and allowed me to use several illustrations. Greg Knapp supported me as department chair. Barbara Parmenter made helpful comments, as did Ian Manners. I thank the Geography Department staff, Sakena Slitine, Kelly Hobbs, Greg Osburn, Patricia Schaub, and Maria Acosta, for their assistance in preparing the manuscript. Ray Sanders helped with illustrations. Bob Holz, Juanita Sundberg, and Andrew Miles supplied illustrations.

At the Center for American Places, I wish to thank George F. Thompson, president, and Randall B. Jones, publishing liaison and editor, for their valuable assistance with the editorial content and presentation of the book. A survey of some of the issues I expand upon in this book appeared earlier in *Ecumene* (3, no. 2 [1996]).

The Eucalyptus

The Geography of Eucalypts

> The importance of forest plantations in meeting the needs for future wood supplies as well as providing other services is well recognized, but few issues are more contentious than that of the role of species of the genus *Eucalyptus* in forestry plantation programmes. Sometimes it seems, from the extreme statements that are made in its condemnation or defence, that eucalypts are either loved or hated.
>
> —DAVID HARCHARIK,
> assistant director-general,
> Forestry Department, FAO, 1995

It has been little more than two hundred years since the founders of modern botany recognized *Eucalyptus* as an evergreen hardwood genus belonging to the myrtle family (Myrtaceae). It is a large genus, consisting of at least six hundred species, almost all of which grow in a mosaic of temperate, desert, tropical, and sub-alpine environments in Australia and Tasmania (see map).

Distribution

Almost all eucalypts spring up around the maritime edges of Australia, the world's smallest continent and sixth-largest country. Only two species do not occur naturally on the Australian mainland. One of them, the Timor mountain gum *(E. urophylla)*, is found in Timor and on nearby islands in Indonesia. Another one, kamarere *(E. deglupta)*, grows in Papua New Guinea,

1

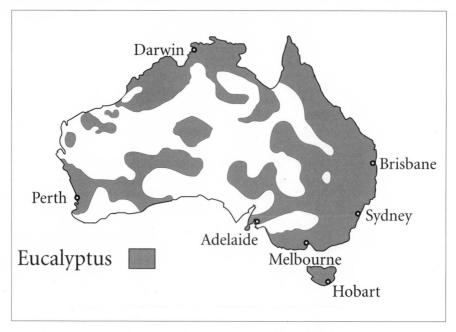

Eucalyptus range in Australia (forest and woodland).
(After Pryor, *Biology of Eucalypts,* and others)

Irian Jaya, the Moluccas, and Mindanao in the southern Philippines. Ten or so species found in northern Australia also crop up in New Guinea (Papua and Irian Jaya), and another species *(E. alba)* occurs in Timor and four islands in the Lesser Sunda group.[1]

Eucalypts dominate the forest flora of Australia and range from 9°N (in the Philippines) to about 44°S latitude in southern Tasmania. Approximately 40 million ha (99 million acres) of native forest—that is, about 5 percent of the land surface—is spread over Australia, mostly along the eastern seaboard, in Tasmania, and in the extreme southwest. Eighty percent of the open forests (forests in which tree crowns shade from 30 percent to 70 percent of the ground area) consists of eucalypts. This plant genus also dominates the broader east and west flanks of the interior, in which trees are widely spaced to form woodland, which together with forests covers a little over 100 million ha (247 million acres), or 14 percent of the nation's land area.

Most of these Australian hardwoods are distributed unevenly. Many eucalypt species are widely dispersed, being more abundant on the more humid coastal fringes, and less common in the arid interior zone, where they favor riparian sites and moist situations. Generally, species are listed as the "summer rainfall group" in northern areas and the "winter rainfall group" in the south. Some eucalypts along the eastern coast have adapted to both regimes; an additional group grows in southwest Australia, where genetic isolation has favored a high degree of endemism. Most eucalypts, and the best of the natural stands, are situated south of the Tropic of Capricorn.[2]

Complex patterns of various species flourish around Australia's northern, eastern, and southern margins, probably because of the sensitive links between plants and growing sites. Many eucalypts occur in uniform stands of at most a few hectares in size. Edaphic conditions, including available nitrogen and phosphorus, plus slope, aspect, fire incidence, and frost, condition and limit the occurrence of a particular species. In general, eucalypts grow in Australia on soils that are low in nutrient value, rich in iron and aluminum, and generally acidic. They do not usually thrive on saline lands or heavy clays. They prefer sands and loams with 600–1,000 mm of annual precipitation and do not survive severe or prolonged bouts of sub-freezing temperatures.[3]

Physical Characteristics

Eucalypts are woody plants of various sizes. Some are among the tallest broad-leaved trees in the world, towering 100 m (328 ft) above the ground. For example, the mountain ash *(E. regnans),* which exists in a discontinuous belt across southern Victoria and Tasmania between 37° and 43°S latitude, is the tallest tree in Australia.[4] It is the tallest hardwood in the world, and second overall to the coastal sequoias of California.[5]

Early botanists exaggerated or miscalculated heights, claiming that some individuals reached more than 450 ft (137 m), 100 ft (30.5 m) taller than California's huge coastal redwoods. The tallest specimens, close to 100 m high, are recorded from the southern half of Australia, where overall dimensions and rates of growth also achieve maximum expression.[6]

The tallest eucalypts are typically associated with dense undergrowth in Australia, although the environmental gradient occupied by the genus ranges from dry and open woodland consisting of medium-sized and taller trees through humid and closed forest. Taller trees tend to promote a better developed understory, because more light filters to the forest floor. Generally, eucalyptus leaves are sickle-shaped and pointed, and tend to become smaller as the parent tree ages. In many species the leaves hang vertically, and they may twist to expose the thicker cuticle to the sun, thereby reducing the evaporation of water from the leaf. In this way eucalyptus woodlands, especially ones with taller and older trees, let in more sun and moisture than other forests. Where a heavy, moist understory exists, periodic fires may be required for successful regeneration. This occurs rarely, perhaps once every century or so, when drought makes a natural burn possible.

In drier zones wildfires sear the countryside every few years or at least once a decade. In such woodlands the composition of plants is more varied. Some respond better than others to burning. Older trees, for example, cope with fire through thicker barks that make them relatively immune to flames. Others have the ability to coppice after a burn—that is, they send up shoots from blackened stumps or from lignotubers beneath the ground surface. This characteristic of regeneration after damage applies to some ten or so of the most widely grown eucalypts, which foresters promote for wood production on an industrial scale outside Australia. All these timber-producing species grow well from seed and coppice easily after the main trunk has been cut down. Most of them belong to the subgenus *Symphyomyrtus,* which is the most widely distributed of the three subgenera, forming woodlands and forests throughout the continent. Members of this subgenus tolerate drought, soil pests, and poor substrates, but will also thrive in more fertile situations.

At the other end of the size scale are dwarf or shrublike eucalypts. There are twenty or so short-trunked species, called *mallees,* which send up stems from underground wood. They resemble desert shrubs. Some are valued for aromatic oils, but none is used for lumber or pulp.[7]

Most eucalypts are woodland or forest trees about 10–50 m (33–164 ft) high. One of these, the blue gum (*E. globulus* ssp. *globulus*), is a most popular species that is synonymous with the name "eucalyptus" in many areas of the world and has been intensively studied. This subspecies (generally known as simply *E. globulus*) is one of the most widely planted eucalypts overseas. At home, it grows in disjunct areas of Tasmania and southern Victoria, between 38° and 43°30'S latitude. It does well on a range of substrates, including granites, mudstones, sandstones, and dolerites, but does best on moderately fertile loams rather than on heavy clays or on strongly calcareous soils. As a mid-sized eucalypt, it attains 45–70 m (148–230 ft) in height and prospers under a rainfall regime of 500–1,100 mm per year; outside Australia it adjusts to an even broader range of precipitation and grows even taller. It is sturdily built and has a crown of moderately dense foliage consisting of large, drooping leaves. The bark scales off annually in long, thin strips; as the tree matures, the scaling takes place higher and higher up the trunk.

The blue gum grows from near sea level to about 1,100 m (3,600 ft), with elevation increasing toward the southern portion of its range. Although the species coppices after fire (and after sawing) and withstands strong winds, it tolerates neither drought nor severe or prolonged frosts. Mature trees may survive down to −10°C (notably those from the higher elevations, in Tasmania), but seedlings and young trees cannot withstand such low temperatures.[8]

Other eucalypts in Australia are equipped to deal with more extreme moisture and temperature regimes than the adaptable blue gum. One of them is the snow gum *(E. pauciflora)*, a dwarf twisted tree, not commercially attractive, that grows in montane uplands 1,800 m (5,900 ft) above sea level. The snow gum can withstand wet and cold winter temperatures, like another sub-alpine species, the cider gum *(E. gunnii)*. This species grows on the central plateau around Hobart, Tasmania, and remains a significant item in the nursery trade in England, where larger specimens grow in northern counties. It also inhabits poorly drained alpine uplands and responds to cold-hardening down to about −1°C.[9]

The river red gum *(E. camaldulensis)* flourishes alongside lowland watercourses, where it is subjected to inundation and heavy, wet soils. It occurs in sandy floodplains of seasonally dry water courses throughout most of the continent. Commonly reaching half the height of the blue gum, this 20 m (66 ft) tall red gum is among the most widely distributed eucalypts in Australia. Open woodlands of which this tree is a major component exist everywhere except Tasmania, the Nullarbor Plain, and Eyre Peninsula. It is also one of the most successful eucalypts planted in foreign countries, where people rely on it for fuel, charcoal, poles, and fencing. The wood of the red gum is hard and durable, and its flowers are an excellent source of honey.[10]

The overall capacity of the genus to adapt to a range of biophysical conditions and altitudinal zones has helped make it possible for eucalypts to thrive outside Australia. Over the past century and a half collectors, horticulturists, and forestry scientists belonging to various national and state agencies have successfully promoted the introduction, establishment, and spread of about a hundred species. These trees have done best in the warmer zones of Eurasia, Africa, and Latin America. Released from nutrient-limited or nutrient-deficit situations, the alien plants have achieved spectacular rates of growth, much higher than for conspecifics in their native habitat and, according to researchers, the fastest rates for any trees on earth.[11]

Once established elsewhere, some species of eucalypts are capable of adjusting to a broader range of soil, water, and slope conditions than in Australia. One explanation for this plasticity holds that, once released from interspecific competition and from native insect fauna that have co-evolved with eucalypt speciation, the woody plants have thrived in unexpected places. Another explanation refers to the shifting conditions of flood, drought, and heat to which the genus has been forced to adapt over millions of years. Ancestral species may have evolved in tropical rainforest conditions but were subjected to periodic aridity, cooler seasonal temperatures, shifts in soil characteristics, and geographic isolation. Climatic and tectonic shifts in the geological history of the Tertiary Period in Australia have selected for adaptive qualities, such as resiliency, in its plant cover.

Ross Florence suggests that about 20 million years ago, fires became more common in Australia, accelerating the transition from rainforest to more open forests and woodlands composed of eucalypts. How much the Aboriginal people affected the distribution of the genus through an accelerated use of fire after their arrival more than 30,000 years ago is open to speculation. Florence believes the spread of woodlands was based upon "the progressive evolutionary adaptation of a 'drought tolerant mesophyte' to the declining and increasingly variable rainfall of the present woodland zone." However, other experts argue that more frequent, less intense burnings are responsible for the ubiquity of the genus, and that this current dominance is due to human activities.[12]

Pests and Predators

Entomologists note that a very large number of leaf-eating insects, such as phasmatids (stick insects), chrysomelids (leaf beetles), and scarabs (Christmas beetles), feast on eucalypts in Australia, as do additional sucking and boring species, such as psyllids (lerps), Isoptera (termites), and wood-boring Lepidoptera. But because almost all populations of trees outside Australia have been grown from seed, very few of their insect parasites have traveled with them. Neither, of course, have the Australian mammals, which also limit survival of the trees in native habitats.

Wallabies, opossums, and pandemelons are three of approximately fifty-seven native mammals, together with an additional seventeen species that have been released in Australia, which are known to inhabit tall-crowned and open eucalypt woodlands. (About 234 species of native birds—about half of Australia's terrestrial avifauna—also occupy these wooded environments.) Most of the large animals are marsupials, which browse and defoliate young trees and may control rates of survival, especially after fires have weakened individual stands and insects have moved in. The absence of these animals, as well as the insects, has been important for the success of eucalypts outside Australia. In fact, most wild and domesticated mammals, including the so-called hoofed locust, the domestic goat, will not feed on eucalyptus bark, leaves, or seedlings.[13]

In general, foresters have discovered that switching a species from an area of summer rains within its aboriginal range to a zone of winter rains is more likely to enhance prospects for acclimation in a new locality than the reverse. It is safer to move species to closely similar soil and rainfall regimes, and to avoid moving them to areas with more extremes of cold than in their native habitats. Most trees from North America and Eurasia grow well in Australia, especially southern Australia, but few Australian trees will grow very far north of the equator, since they are adapted to heat, drought, and fire, but not to prolonged or severe cold.[14]

For more than half a century growers have used climate as the rule of thumb for matching eucalypt species with new sites. For example, southeast Australia and Tasmania have temperate humid zone equivalents in New Zealand, western Europe, the northwest United States, southern Chile, and northeast Argentina. Eucalypts may be transported from this part of Australia to these regions, but there are associated risks, as seen in the case of blue gum. Lower minimum temperatures in areas away from the native range may prove lethal to the trees and restrict their adaptability. Foresters have experimented with "hardening off" blue gum—that is, exposing plants to lower and lower temperatures in order to make them better able to cope with severe frosts, but so far such trials have proven unsuccessful.[15]

Better matches are now being made through computer-based analysis using a large number of variables, among which rainfall periodicity, the duration of the dry season, and variations in mean temperatures appear crucial. These technology-driven analyses generate maps of potential new ranges—for example, one for *E. grandis* in sub-Saharan Africa. The map is then compared with the areas in which trees already exist in order to refine the "fit."[16]

Overall, frost appears to be a limiting factor for many eucalypts. Freezing weather can kill trees over thousands of hectares, but it is not yet clear just how low the temperature has to be. Whether specimens survive may well depend on the series of night temperatures that precede a severe freeze. Cold air originating in very cold regions may destroy trees after several nights with cool but nonfreezing temperatures. Eucalypts do not go dormant in winter, and thus the trees may die, as they do in southern France, Florida, or around the Black Sea when Arctic air descends suddenly over plantations unused to clear air freezing.[17]

Oils

Eucalypts are important sources of medicines, industrial oils, and aromatic chemicals used in the manufacture of perfumes and soaps. Tannin may be extracted from the bark of many species whose resins contain kino-tannic acid, used in the manufacture of throat lozenges and mouthwashes. Each species has its own kino, just as its leaves possess essential oils and chemicals in various quantities and combinations. Leaf oils from approximately twenty species have medicinal properties and are turned into pharmaceutical products, such as antiseptics, disinfectants, decongestants, diuretics, stimulants, and balms. The essential oils are extracted from fresh leaves through the process of steam distillation.[18]

Eucalyptus oil was among the very first natural products shipped from Australia. In November 1789 Denis Considen, assistant to the surgeon-general of the First Fleet, John White, filled and dispatched a bottle of oil to Sir Joseph Banks on behalf of Arthur Phillip, first governor of Australia. White obtained the oil from the so-called peppermint tree, which reminded him of *Mentha piperita* back in England, and declared that it relieved "all cholicky complaints." Chemicals from *E. piperita* consist of phellandrine, piperitone, and cineol— components identified some fifty years later when eucalypt expert Ferdinand von Mueller distilled leaves and branchlets. In 1854, at von Mueller's suggestion, his pharmacist friend Joseph Bosisto established the first factory for the commercial distillation of eucalyptus leaves close to Melbourne, Victoria. Over the following thirty-seven years, Bosisto displayed his chemical extracts at seventeen international exhibitions, making eucalypt oils famous.[19]

By the 1870s the leaves and smaller twigs of *Eucalyptus globulus* had become a major source of oil used in preservatives, disinfectants, deodorants, and sedatives. Blue gum oil from natural woodlands in Tasmania became a regular article of trade in western Europe. However, the oil content of its leaves does not match that of some other species. For example, leaves of *E. polybractea*, blue mallee, which became a mainstay of the Australian oil industry, yield three or four times more oil than blue gum's eucalyptol or cineol, oil that is used in the treatment of colds, congestion, and disorders of the skin. Oils from other gums, such as *E. smithii, E. radiata,* and *E. cinerea,* are processed into stain removers, toothpaste, and toiletries. Sachets of citronella distilled from the lemon-scented gum *(E. citriodora)* are desirable as insect repellents. And chemicals from three additional species go into perfumes and food flavorings. Peppermint-smelling leaves of *E. piperita* and its variants supply piperitone and phellandrene-rich oil used in industrial solvents and as flotation for zinc and lead. This substance is also the raw material for the manufacture of synthetic thymol and menthol (from *E. dives*) that goes into cigarettes.[20]

A tradition of folk and herbal medicine continues to involve the use of eucalyptus. Cured leaves may be smoked against asthma, and inhalations are said to help respiratory ailments by opening up the bronchial passages. Oil extracts are used to suppress bacterial infections and skin parasites, and also used in uro-genital complaints. Century-old reports from California note the popularity of poultices and teas made from leaves. Steam wafted through new leaves were believed to help suppress colds and fevers. A few oil drops in bathwater were thought to relax both the body and mind. Other preparations went into candies, shampoos, and tobacco—all reflecting the high standing of eucalypts, particularly blue gum, "among the laity," as one expert concluded.[21]

Today, aromatherapists single out eucalyptus for various remedies. The colorless mobile liquid with a camphorous odor and wood-sweet tone mixes with rosemary, lavender, marjoram, pine, and lemon to produce concoctions that are intended to enhance circulation, respiration, skin care, and the wellbeing of the body's nervous and immune systems.[22]

Honey

Australian researchers have attested to the "mellaginous" character of native eucalypts, which blossom in every month of the year. André Métro identified approximately thirty-five species valued for their pollen, nectar, and honey. He listed the finest honey producers as river red gum, tuart *(E. gomphocephala), E. melliodora* (whose honey "is the finest in Victoria"), Sydney blue gum *(E. saligna),* forest red gum *(E. tereticornis),* and messmate *(E. obliqua).* This secondary product has helped enlist and bolster regard for eucalypts outside Australia.[23]

Norman Ingham, foreman of the University of California's Forestry Station in Santa Monica, noted the role of flowering eucalypts as "bee pasture" in southern California at the turn of the twentieth century. The insects appeared to prefer whitish or greenish white flowers as they foraged among the trees. He calculated that the insects were able to visit the flowers of at least four to six species in the arboretum during most of the year.[24]

Nomenclature

Eucalypts have unusual names. Many refer to the grain, color, and hardness of the bark and wood. Silviculturists commonly employ six general names that refer to the various textures and colors of the bark: *boxlike, wrinkled, stringy, ironlike, scaly,* and *smooth.* The names for eucalypts also include characteristics of potential timber and serve to divide eucalyptus trees into broad classes. *Ironbarks,* for instance, grow with dark, hard, thick, fissured or corrugated trunks, impregnated with resin. *Stringybarks* have fibrous barks that peel away in long pieces. *Boxes* have hard, quite thick, tightly woven barks that are not as corrugated as ironbarks. *Peppermints* possess fine fibrous barks rather like stringybarks and put out leaves full of various eucalypt oils with a pungent smell and taste, some similar to peppermint. Finally, *gums* usually possess smooth, thinnish, often light-colored trunks, although some experts apply that name to those species whose barks flake off in long ribbons or flakes.[25]

Gum has another meaning. Explorer-botanists frequently used this term to describe resinous secretions from both the trunk and leaves. The term has been attributed to Captain Arthur Phillip, founder of the Australian settlement, who noted in May 1788 that "what seeds could be collected are sent to Sir Joseph Banks, as likewise the red gum taken from the large gum tree by tapping." In Australia, the term often refers to the entire genus.[26]

This group is of particular importance because the gum trees, notably blue gum, "a jack-of-all-trades tree," are the best known representatives of the genus outside Australia. However, this common name can also be misleading. *Blue gum* applies not only to *E. globulus* in Tasmania (also called Tasmanian or southern blue gum), but also to *E. tereticornis* in Queensland, to *E. leucoxy-*

lon in South Australia, and to *E. saligna* in New South Wales. Even in New Zealand the imported *E. biangularis* is frequently called blue gum.[27]

It is useful to regard eucalypt species as discrete taxa, but there is a continuum of variation not merely between species but also within a single species. Even the bark of a particular eucalypt may change according to soil and moisture conditions in various places. Subdivisions are also complicated by the large number of species within the genus, and the wide distribution of the genus throughout Australia, necessitating comparisons in order to separate out allied plants.

Efforts to classify eucalypts commenced with a study, undertaken by L'Héritier in the late eighteenth century, of the size and shape of the operculum, or cap, over the flowers, which is peculiar to eucalypts. Research in the nineteenth century shifted to leaves and leaf-stalks and to variations in barks, before settling upon the size and shape of the anthers, or pollen-bearing lobes, in the flowers. Additional efforts to subdivide eucalypts have also included exudation of essential oils, fruits, and qualities of timbers.

In a revision of the plant genus published in 1909, Australia-based botanist Joseph Henry Maiden (1859–1925) concluded that among plant scientists, the flower buds and fruits had proven most useful in classifications, whereas among foresters, characteristics of barks and timber were paramount. Recent research findings have stressed genetic features resulting from geographic isolation and hybridization. Some species have distinct and constant characteristics by which experts may recognize them, others do not. In the problematic cases, identification may include several factors, such as the tree's overall shape and size, its bark, oil glands in the leaves, flower size, and seedlings, as well as the location and habitat in which a specimen is growing. Recently, botanists have made a good case for reclassifying eucalypts into several genera, although some foresters and applied scientists regard this as an unacceptable step toward confusion.[28]

Increasingly, the standard common or vernacular names are falling into disuse, and more and more foresters are referring to them by their specific names; thus blue gum becomes *globulus* and not *tereticornis,* which also carries the name of blue gum in northern Australia. The practical aspects of naming eucalypts clearly become important when one considers that documented instances exist of commercial orders for a particular species being filled by seeds from another due to a mix-up in their common names.

Rural Australia

Over the past two hundred years, European colonists and their descendants have cleared native eucalypts to make room for crops and pastures. The trees are the backdrop for contemporary settlement, although ranching on the continent's coastal rim has altered the distribution patterns of many forest and woodland species.

Eucalypts anchor the steeper slopes and provide shade when intermingled with introduced conifers in fields and around farmsteads, as here in New South Wales.

Old iceboxes and refrigerators used as post and delivery boxes near Tarago, New South Wales, are reminders of ranching, which cleared eucalypts to form new pastures.

Eucalypt Adaptability

Australia has more than twenty-five hundred tree species, including some seven hundred eucalypts. The diversity of the continent's flora, its extensive distribution, high intraspecific variation, and adjustment to a range of environmental factors make it attractive for uses overseas.

Wind-sheared and frost-sculptured trees on the Brindbella section of the Great Dividing Range, along the border of Australian Capital Territory.

Fog-shrouded montane woodland on the Brindbella Range, Australian Capital Territory.

Pindan country of *E. dampieri* woodlands on red sands, located on the remote Dampier Peninsula, Western Australia.

Eucalypts, wattle, and termite mounds fringe the Roebuck Plains south of Broome, Western Australia.

Eucalypts form heaths along maritime headlands, as at Mimosa Rocks, New South Wales.

Galleria forests line inlets and estuaries. This tall stand is beside a milking shed at Wapengo Lake, New South Wales.

Riparian sites, such as this one on the
Murrumbidgee River, near Tharwa, New
South Wales, show a range of trees and
shrubs.

Large eucalypt on Lake George, near
Bungendore, New South Wales.

Urban Eucalypts

Eucalypts are an integral part of city vegetation in Australia. There is a movement away from introduced trees and toward planting native species in urban parks and suburban areas, such as Canberra, the nation's capital.

Eucalypts are important landscape elements on the campus of the Australian National University, Canberra, and are well represented in the adjacent botanic gardens.

Eucalypts, including *E. citriodora,* grow among vineyards in Rutherglen, New South Wales.

Bound for Europe

Eucalypts for Pleasure

> On account of their leaves, flowers, bark,
> scent, and habit, all so unlike those of any
> other European trees, they form an attractive
> feature in gardens and pleasure grounds;
> and are so easy to raise from seed, that the
> certainty of their death after a few years will
> not deter gardeners from planting them.
>
> —HENRY ELWES AND
> AUGUSTINE HENRY, 1912

Discovery

Plant collectors in the Southern Hemisphere, specifically Oceania, brought the genus *Eucalyptus* to world attention. In spring 1770 Joseph Banks (1743–1820), who later became director of the famous Royal Botanic Gardens at Kew, and his friend and assistant Daniel Carl Solander (1733–82), whose mentor was Swedish taxonomist Carl von Linné (Linnaeus), collected the first specimens for transit to Europe. They made this botanical discovery as members of a British Royal Navy expedition during Captain James Cook's first voyage to the Pacific Ocean (1768–71). Admiralty instructions had directed Cook to Tahiti ostensibly to observe the passage of the planet Venus across the face of the sun. After making these astronomical observations, Cook steered the *Endeavour* westward. Nine days after spying the coast of "Terra Australis" (present-day Point Hicks, Victoria), whose coastline the captain likened to "the back of a lean Cow," the ship hove to not far from present-day Sydney. It was there on 1 May 1770 that Banks and Solander encountered their first eucalyptus trees.

In his personal log, Captain Cook noted that "the great quantity of NEW PLANTS ETC. Mr. Banks and Dr. Solander collected in this place occasioned my giving it the name of Botany Bay." Cook anchored in what he had at first called "Stingray Bay" for eight days, and on a twelve-member excursion into the hinterland with its sandy soils, grasses, and scrub, Cook and his crew from the *Endeavour* encountered "very few species of trees, one of which," Banks recalled, "was large yielding a quantity of gum much like *sanguis draconis*." The explorers mistakenly identified eucalyptus sap as exudate of the dragon tree from the Canary Islands and Madeira. The trees they encountered, however, were eucalypts, probably belonging to the species *E. gummifera*. London's Linnean Society possesses a eucalypt attributed to that first expedition, which "probably languished undescribed and unnamed."[1]

Banks noted eucalypts on at least two other occasions as the vessel sailed northward along the coast of New South Wales in the late summer and fall of 1770. At another landing he described other gum-bearing trees, which "differed however from those seen in the last harbour in having their leaves longer and hanging down like those of the weeping willow." These were probably *E. crebra*, narrow-leafed red ironbarks, which he mentioned at another landfall as having ant nests among their branches.[2]

Eighteen years later, at the behest of Banks (who was then in charge of Kew and shortly to become a baronet), Charles Louis L'Héritier de Brutelle (1746–1800), a French plant specialist, was working in London on materials collected at Adventure Bay during Cook's ill-fated third expedition (1776–79). Self-taught scientist and ardent follower of the Linnean system, L'Héritier proposed the generic name *Eucalyptus*. His specimen was a dried plant pressed by the expedition's surgeon William Anderson on Bruny Island off the southeast coast of Tasmania in January 1777. Having served on Cook's second expedition to the Pacific Ocean (1772–75), Anderson was familiar with the fragrant plants, which he named "Aromadendron." Like his commander, Anderson died on the final expedition, but his Kew-based assistant David Nelson safeguarded the plant materials on which L'Héritier subsequently worked.

L'Héritier's name for the genus comes from a Greek root *eu* for "well" and *calyptos,* meaning "covered." *Eucalyptus* refers to the seal, or cuplike operculum, that covers and protects the flowers. The second part of the name of his particular specimen, *obliqua*, refers to the unbalanced growth of the leaf base of this particular species. Today its common name is messmate in Victoria or stringybark in Tasmania, after the texture of the thick, fibrous, brown bark. L'Héritier's book *Sertum Anglicum,* published in Paris in 1788, consisted of thirty-six pages and thirty-five plates and described thirteen new genera of plants, including eucalyptus.[3]

Curiously, the same species that provided the dried specimen upon which L'Héritier worked also grew as a small tree in the Royal Botanic Gardens at Kew, thanks to the initiative of Tobias Furneaux (1735–81), a Royal Navy officer who commanded the second ship of Cook's 1772 expedition to the Southern Seas. In February 1773 Furneaux's vessel had become separated from that of his commander, so he had set course for a predesignated rendezvous with Cook at Queen Charlotte Sound, New Zealand. En route he decided to put ashore on Bruny Island, Tasmania, on 10 March 1773, almost four years before Anderson and Nelson landed there.

Laying up for five days to draw water, gather wood, and overhaul the *Adventure,* Furneaux and his crew explored the hinterland, noting among other features some tall evergreen trees whose leaves were long and narrow. He recorded that "the seed (of which I got a few) is in the shape of a button and has a very agreeable smell." The head gardener at Kew, William Aiton, planted one or more of these "buttons" that were ferried back to England in July 1774, from that chance landing. A live *E. obliqua* specimen is recorded in the official index *Hortus Kewensis* in 1789.

Cook's men, however, were not the first Europeans to observe the gum tree family. There are tales that Portuguese explorers or Dutch traders collected eucalypt seeds and residues. Eucalyptologist Max Jacobs, for example, speculated that Portuguese seamen carried away seeds of eucalypts from Timor in the 1520s and could have introduced the native *E. alba* and *E. urophylla* to Brazil before that nation's independence in 1822. Jacobs noted that Brazil was a usual stopover for vessels sailing from Australia to Europe via the Cape of Good Hope.[4]

In 1688, almost a century before Cook's voyages, another English explorer, William Dampier (1651–1715), had noted gumlike extrusions from trees he spied in an anchorage in northwest Australia. But this pirate, adventurer, and Royal Navy officer confused them with the dragon tree, as did Banks.[5]

Success as Exotics

Building upon L'Héritier's initial description, botanists have since studied the genus most closely in Australia and in those foreign countries, including India, South Africa, and Brazil, in which it has achieved economic importance. Currently, more than 550 named species dominate Australia's forest cover, and several dozen have come to be of commercial value in tropical and subtropical parts of the globe. A dozen or so eucalypts in particular are now major components in forestry research and the development of industrial woods outside Australia (see appendix 1).[6]

By the early 1980s industrial stands of eucalyptus existed in approximately a hundred nations. Wooded areas spanned more than 4 million ha (10 million acres), a fivefold increase in twenty-five years. Most eucalypts exist in plantations between 27° of latitude north and south of the equator. By the early 1990s the reported increase outside Australia totaled 200,000 ha (494,000 acres) annually, resulting in a combined world estimate, including plantations and woodlots in the tropics and hotter subtropics, of approximately 16 million ha (39.5 million acres) (see appendix 2, table 1).[7]

Taken together, this expanse totals 61,815 sq mi and is larger than England and Wales together. It is about equal to the area of the state of Washington in the United States. Put another way, the land area taken up by eucalypts *outside* Australia is 10 percent of that covered by natural stands of eucalypts *inside* that continent; the outside percentage is growing steadily.

Today, about 37.5 percent of all tropical forest plantations, which blanket 42 million ha (104 million acres) around the world, consist of eucalypts, compared with plantations in Australia, which remain in their infancy, totaling only 1.1 million ha (2.7 million acres), of which eucalypts comprise only about 200,000 ha (494,000 acres). Currently, this percentage is climbing in the tropics and hotter subtropics as eucalypts are planted as a short-rotation crop. This number will soon exceed 50 percent due to the growing importance of eucalypts in the world production of wood pulp.[8]

A dozen or so species are especially useful in the manufacture of newsprint and other paper. Many of them grow more rapidly and yield more wood in a shorter time span than conifers, which have traditionally supplied pulp in temperate regions. Today, eucalypts are setting records for wood output in the warmer areas of the world.[9]

Early Interest

Aristocrats, wealthy merchants, and other landed gentlemen, who became fascinated with rare or unusual plants, together with the collectors, hor-

ticulturists, and nurserymen who grew and sold them, proved energetic promoters and popularizers of eucalyptus trees. The initial impulse to import,
cultivate, and redistribute members of this new woody genus arose in Europe,
where botanizers sought out new ornamental and useful plants. In England,
Scotland, and southern Ireland, from the 1820s through the 1840s, prominent
persons, through nurserymen, gardeners, and the like, conducted trials with
more than a score of species in order to determine whether conditions, particularly winters, would overwhelm the Australian trees. As the years ticked by,
reports about successes and failures appeared with increasing frequency in
popular books and periodicals devoted to horticulture, gardening, and landscape design.

An organized tradition of collecting and holding trials for native and
foreign plants in the British Isles goes back to 1621 when Henry Danvers, earl
of Danby, gave 5,000 pounds sterling to the University of Oxford to set up a
"physic garden" for the study of drug plants, commonly called simples. His
herb garden was located on the grounds of an old cemetery on the banks of
the River Cherwell opposite Magdalen College, which retains the lease with
the University of Oxford. Within twenty-five years at least sixteen hundred
different kinds of plants were growing in what has been called, for the past 160
years, the University of Oxford Botanical Garden. Other botanical gardens followed, including one associated with the University of Edinburgh (1670),
Chelsea (1673), and the Royal Botanical Gardens at Kew (1759). Under the able
direction of Sir Joseph Banks, Kew grew into a major nursery and research
center, as he and his successors sent collectors to South Africa (1772–73, 1786–
96), the Canaries and Azores (1778–82), and Spain and Portugal (1783–85), in
order to establish collections of living plants. When L'Héritier described *Eucalyptus,* approximately fifty-five hundred species of plants were already growing on the botanical grounds.

Commercial nurseries in and around the capital also contributed to
plant acquisition and study. The most renowned was Brompton Park Nursery,
established in 1681, which stocked plants that had been shipped into the nation
over the previous three hundred years. In 1700 Brompton covered more than
a hundred acres and employed twenty-two people who tended stock valued at
40,000 pounds sterling. Another important horticultural outlet was one in
Hackney, founded by a German, John Busch, and purchased by a Dutchman,
Conrad Loddiges, in the early 1770s. Loddiges & Sons was famous for its orchids, but advertised more than thirty eucalypts and other species from Aus-

tralia and New Zealand among two thousand greenhouse plants in the catalogue for 1830.[10]

Experiments in the British Isles

In the first decades of the 1800s John Claudius Loudon, an inventor, architect, and landscape planner, took an avid interest in gardens and in growing all sorts of plants. In an exhaustive survey he published about trees and shrubs in Britain, Loudon declared the new *Eucalyptus* genus to be "a very remarkable one," and anticipated that "proprietors in the South of England would encourage their gardeners to plant out these, and other Australian trees, in dry sheltered places in their shrubberies and woods." Seeds were reasonably inexpensive, as were small trees (one could purchase one in Scotland for a shilling), and the eucalypts with which he was familiar seemed quite hardy. With a hint of optimism, Loudon noted that the trees did best in the home counties and in the south and southwest.

Appropriate trials were necessary, stated Loudon, in order to assess thoroughly the overall tolerance of eucalypts to prolonged cold and frosts. Although he concluded that British summers were not hot enough to "ripen" the wood for timber, and the winters were in general too cold for most plants to be grown widely, Loudon speculated that "probably most of the species, if planted so as to form one entire wood, would protect one another." His imaginary wood was to grow somewhere in southern England. Even if eucalypts did not develop into timber-sized trees, in his view, they would be kept alive as some form of dense copse. Presumably, this cover would be a feature in which well-heeled entrepreneurs, such as himself, could indulge a passion for hunting.[11]

Loudon noted that eucalyptus specimens had surprised growers since the initial introduction of the genus into England in about 1788, the year of initial English colonization of Australia. One tree at Saxmundham, Suffolk, planted by the wife of Sir James Edward Smith, a famous author and plant expert, was 20 ft high (6.1 m), with two stems "each as thick as a man's leg." Around the capital city "it requires very little protection when planted against a wall," Loudon advised, deciding that the stringybark was definitely able to withstand "the open air in mild winters." Loudon knew of at least one swamp mahogany *(E. robusta)* in the Horticultural Society's botanic garden in London, and another one on Stamford Hill. Other eucalypts grew in the same locations, or on similar sites; a ten-year-old tree, which had been planted in 1825, he noted, had begun to flower in the Royal Botanic Gardens at Kew.[12]

In the 1830s, when Loudon penned his remarks about these new plants, a growing number of eucalypts were prospering in England (though what species they were was not always clear). Hardiest specimens, according to expert testimony, came from Tasmania. In a summary published in 1832, Loudon listed fifty-six species in Britain. All of them, he noted, needed to be raised in the greenhouse, and almost two-thirds he labeled as ornamental rather than wood-bearing trees. Between 1788 and 1800, this author reported five species arriving for the first time; during the following decade an additional fourteen became established. Interestingly, only one of the first fourteen species on his list was introduced for ornamental purposes; most entrepreneurs, it seemed, regarded eucalypts as potential wood-producers. Understandably, usefulness took precedence over good looks, at least initially.

But expectations soon changed. Loudon classified as ornamental almost all of an additional fourteen eucalypt species that arrived between 1810 and 1820. And in 1823, when he concluded his summary, only two of the sixteen new species were expected to yield wood of any value.[13]

Correspondence published in the London-based weekly *Gardener's Magazine* bolstered Loudon's judgments about cold-hardiness. "In Britain they are generally kept in a green-house or pit," noted an 1830 entry. Though "hardier than the common myrtle," and growing "freely on any soil," they are killed to the ground by frost, but around London "seldom fail to spring up again."[14]

During the 1830s reports from Norwich and as far north as York supported the belief that seed originating from Tasmania, notably from around Hobart, turned into the most resilient specimens.[15] Also, news from abroad, from Caserta and Naples, Italy, and near Funchal, on the island of Madeira, reported that eucalypts had already grown upward of 60 ft (18.3 m) high. Expectations were being realized.[16]

In England, eucalypts rarely achieved such heights, due to freezing weather and high winds. Temperatures in January 1838, for example, fell to 22°C and killed off many trees. A year earlier, a similar big freeze at Hyères, near Nice, in southern France had hit them hard.[17] Throughout the remainder of the century and into the 1900s, hard freezes (1853/4, 1860/1, 1880/1 [which destroyed all the blue gums on the Scottish mainland], 1894/5, and 1908/9), plus storms and gales (for example, in Cornwall in 1859) splintered or toppled these speedy growers. The tallest blue gum, which stood 33.5 m (110 ft) high on Rozel Bay, Jersey, died in the winter of 1894/5; the oldest, at Tresco on the Scilly Isles, lost its top branches several times until it eventually blew down in 1891 at the age of forty-one. Trees often raced upward without a firm and steady grip on the soil. In this way, cold and windy weather limited the distribution and numbers of eucalypts in the British Isles.

Winter and storm-related problems at home and on the European continent killed or stunted eucalyptus trees once every decade or so and eventually soured English growers on raising them. But in the first period of trials and establishment, observers eagerly set out seeds and lavished attention on the saplings. Some promoters used the popular media, such as *Gardener's Magazine* and *The Garden,* to instruct colleagues about how and where to grow these botanical novelties, informing growers of the need to select warm, sheltered sites and to avoid exposed places. Devon, Cornwall, and the southern counties in Ireland appeared optimal locales. One correspondent from Derbyshire, a marginal area, even fashioned "a thick carpet of hair" around the base of his blue gum to insulate the roots of his precious tree.[18]

The reasons for planting eucalypts varied. Some buyers enjoyed the novelty and challenge of raising a rare and little-known plant. Others found that tending eucalypts successfully for the conservatory or garden gave them a sense of accomplishment and satisfied their curiosity. People who had encountered larger specimens or had observed eucalypts in their native habitats agreed that the genus was likely to be useful and attractive, combining graceful boughs with variously textured trunks and soft-colored leaves.

Gardener and landscapist William Robinson, famous for strongly worded opinions about ornamental plants and landscape gardens, had surprisingly kind words for "the handsome Australian trees," as he called them. He admired their bearing and the leathery texture of their leaves, "always entire and very variable in their shape." As soon as they grew to 12 ft (3.7 m) or so high, they assumed for him a distinct and graceful habit. Robinson was struck in particular, he recalled, by the stature of lofty specimens in California, although English growers would have to be content to read about such impressive eucalypts.[19]

Other admirers echoed Robinson's views. Eucalypts added distinctive hues and shapes in a large garden or estate, and by interspersing them with other plants gardeners and landscapists created a variegated panorama of trees and shrubs. Madame Tzikos de St. Léger, who grew nineteen peppermint gums at her island home in Lago Maggiore on the Swiss-Italian border, admired the "white trunks and light green leaves against darker forms of conifers behind." Moving beneath them, she was able to brush against their bark, which hung in strips, and thereby imagined herself "walking under the lianas of a virgin forest."[20]

A visitor to an estate in Algeria had a similar experience: "[It] produced a curious impression to walk in the dim twilight of this Australian-African forest, and to think that that this was also a wood of the Miocene period. Beautiful is not the word I should apply to its appearance; but in exchange for sun-baked earth and deadly swamps, I must say these eucalyptus forests are most grateful, and the smell delightfully resinous, warm and gummy."[21]

A 20 ft tall (6.1 m) manna gum *(E. viminalis,* a species extensively distributed in southeast Australia), positioned in a bank of hardy palms backed by conifers, reminded another correspondent of "a column of blue smoke." Many growers found this bluish tinge in the foliage of young blue gums singularly appealing, although as the trees mature the tint tends to fade.

The unusual scent of eucalyptus also drew high praise. The combined effect of fragrant gum blossoms bursting from pendulous leaves on sloping limbs persuaded many of its ornamental merits. Toward the end of the century the Paris flower market sold petal-laden branches of yellow box *(E. melliodora),* swamp mahogany *(E. robusta),* blue gum *(E. globulus),* and yellow gum *(E. leucoxylon),* as well as one or two other species, as vase ornaments suited for drawing-room mantels and windowsills. Suppliers shipped them from warm, sun-dappled woodlands that dotted the French Riviera.[22]

Back in England trials continued, succeeding satisfactorily only in the warmest regions. The Isle of Wight was one balmy place where at least three species achieved some stature. One reached a height of 8.5 m (28 ft) in barely four years ending in 1843. However, a column in the *Gardener's Chronicle* suggested ominously that another, smaller plant, which had germinated from seed supplied by a Neapolitan botanist, seemed inclined "to refuse our winter." In March 1844 eight or ten eucalypts in Northwood Park, Cowes, had survived the coldest months unscathed and were "doing better than usual," declared an advocate.

Other species, mostly red mahogany *(E. resinifera),* widely distributed along Australia's east coast, sprouted equally well several years later on the nearby mainland. Individual plants survived in Exeter (in James Veitch's establishment at Mount Radford, Topham Road, where one flowered) and at Abbotsbury in Dorset. There, a specimen grew to 5.5 m (18 ft) in sixteen years.

This latter mention probably refers to the Ilchester family's famous estate where about forty species of gum trees received trials through the early 1900s. Two additional "luxuriant" specimens (probably *E. globulus*) of similar size grew in John Luscombe's seat at Combe Royal, near Kingsbridge in Devon.[23]

Coastal sites in Scotland and Ireland that were warmed by the Gulf Stream also proved to be excellent homes for eucalypts, at least for a while. In the late 1840s in Scotland, James Maitland Balfour planted seeds of what has proved consistently to be among the most hardy eucalyptus, the cider gum *(E. gunnii)*, endemic to Tasmania, or a hybrid of this species. He reportedly obtained the seeds from the marquis of Salisbury, who had carried them from Tasmania. Twenty years or so later his son Arthur James Balfour, who became prime minister of Great Britain, also visited Australia and reportedly returned with eucalyptus seed, which he planted at the family estate of Whittinghame House, East Lothian; he concluded that the cool, moist, gravelly conditions 400 ft (122 m) above sea level suited their temperament.

The first cider gum hybrid, the so-called *E. whittinghameii primus,* reached a height of 70 ft (21.3 m) and a girth of more than 12 ft (3.7 m). It survived a severe frost in 1860/1. Subsequently, its shoots lost all their leaves in another wintry blast in 1894/5. But its progeny, the so-called *E. whittinghameii secundus,* was distinguished as the first eucalypt to be raised in Britain from home-grown seed (planted in 1885). The estate's gardener reported that during 1894/5, this 20 ft (6.1 m) high *secundus* tree had "not lost a leaf, though the temperature was twice at zero." A dried specimen from the Whittinghame eucalypt is in the collection at Kew Gardens, labeled as being from East Lothian, 1888. Because of its hardiness, the "Whittinghame hybrid," according to tree experts Henry Elwes and Augustine Henry, was the only eucalypt to have succeeded outdoors at Kew (through the early years of the 1900s). Hundreds more of this cider gum hybrid were raised and spread throughout the United Kingdom after about 1887.[24]

Suggestions from travelers familiar with the genus in Australia, together with the practical details supplied by professional horticulturists, notably Baron von Mueller in Australia and Sir Joseph Hooker in England, helped direct trials and enliven popular support for establishing a greater profusion of eucalypts. At a meeting of the Edinburgh Botanical Society in 1876, for example, Rev. D. Landsborough noted an experiment for growing several species on the Isle of Arran in the Firth of Clyde. A blue gum had grown a foot in the first year, then shot up four feet in the second, and another six in the third. The clergyman declared that he expected "to see it generally introduced in a few years, when it will form a valuable addition to our evergreen shrubs."[25]

Showcasing Eucalypts

Special events bolstered support for the plant genus by demonstrating that eucalypts could be hewn into valuable timber. At the famous Crystal Palace Exhibition in 1851, "two huge blocks" of *Eucalyptus globulus,* shipped from Tasmania by Sir William Dennison, went on display. The Great Exhibition, as it was called, drew 4.2 million visitors to 20 acres (8 ha) of convention space from 1 April through 30 September. The section on colonial produce, in which these massive pieces of wood stood out, drew attention to the immense size of eucalypts in their native habitat and also to their considerable value as lumber.

Already, declared one witness, both the secretary of state for the colonies and the lords commissioners of the Admiralty were recommending the use of blue gum wood, whose planks "surpass those of any other timber." The blocks of prime wood in the exhibition pointed up the usefulness of regions about which most people knew little. Eucalyptus and other natural resources proved that colonies were a treasure-trove of raw materials that could help ensure Britain's success against its rivals. All in all, this display of products from far-off lands reassured the British public about the well-being and health of "distant friends," as one reporter put it. He saw "no reason why these prodigious trees should not, at some future day, decorate the scenery of Great Britain. Devonshire and Cornwall, or Cork and Kerry, would certainly prove capable of bringing them to maturity."[26]

Eucalypts valued as timber figured in later exhibitions and fairs. They were shown off in 1855 at the Exposition Universelle in Paris, for example, and in another London show, the Colonial and Indian Exhibition, in 1886. In the latter display, each colony had a court for its plants and animals and its manufactured goods. Gum trees were featured as valuable timber in the displays from New South Wales (in which thirty-seven eucalyptus species were listed), Victoria (twenty-four species), South Australia (ten species), Queensland (forty-seven species), and Western Australia (eight species).

Two upstanding 15 m (50 ft) logs of karri *(E. diversicolor),* showed off the tall, straight trunks and fine-grained wood of that tallest tree of Western Australia. A set of round logs of jarrah (another Aboriginal name; *E. marginata*), consisting of 148 cu ft (4.2 m³) of lumber and weighing almost 5 tons (5,000 kg), served as the portal for that state's court. Both eucalypts produced first-class sawn timber. After the event, both karri and jarrah specimens were secured for the Royal Gardens and displayed in Museum No. 3 at Kew. Jarrah, from southwest Western Australia, drew interest as a possible new surface for streets. In May 1896 a "Special Committee appointed to Consider the subject of Wood Paving in the Parish of Paddington" recommended that durable jarrah wood replace commonly used softwoods. A note on "Wood Paving at the West End" in the *Daily News* of 17 August 1897 confirmed this opinion by noting that the vestry of Paddington had gone ahead and borrowed 13,000 pounds sterling from London County Council in order to surface about 8 mi (13 km) of city streets with this durable hardwood. A total of 850,000 blocks were to be supplied at 10 pounds, 17 shillings, and 6 pence per thousand. Nowadays the production area of this fine timber has been reduced to 1 million–2 million ha (2.5 million–5 million acres) through felling and diebacks.[27]

Exhibitions and fairs drew attention to other gum species and to the products they supplied. In addition to important ones in London and Paris, displays in Australia (in Melbourne, 1861, 1866, 1880, and 1888 [the Centennial]; in Sydney, 1870, 1879; in Adelaide, 1887 [the Jubilee]) and in major cities such as Dublin (1865), Philadelphia (1876), Amsterdam (1883), and Calcutta (1884) served to expand and promote interest among government agencies and the general public about Australia's resources and manufactured goods, especially timber and oils. Exposition catalogues explained how eucalyptus wood was well liked by engineers, architects, wheelwrights, and shipbuilders, and on a more selective basis suited for cabinetmakers and woodcarvers.[28]

There was some concern over the blue gum, whose timber admittedly warped, shrank, blistered, and split as it seasoned after cutting. However, those defects never surfaced in the line of wood products from South Africa mentioned in the 1886 Colonial Exposition. From its arrival in Cape Colony in 1828, it was explained, blue gum had done especially well, exceeding all claims for it back in Australia.[29]

Official enthusiasm had its counterbalance, however. "Few plants have been the cause of more disappointment than the Blue Gum," complained a leading article in the *Kew Bulletin* in 1903. The review essay concluded, in the words of a University of California forester, Professor Hilgard, that *E. globulus* was fit for "firewood only." Wholesale plantings had failed to realize useful timber in the United Kingdom, or for that matter in many other areas, and given their susceptibility to cold (anything below 12°F [–11°C] killed them), only the "juvenile form" appeared in English gardens. It was "a decorative plant in summer" in the warmest portions of the south coast. Then the essay shifted from wood qualities to an exploration of the reputed benefits to human health from growing this and similar eucalypts.[30]

Health and Disease

In the late 1860s through the 1870s, promoters discovered important new characteristics of eucalypts. They learned that the roots of fast growing eucalypts dried out wet and swampy soils, while fragrant aromas from their leaves "cleansed" the atmosphere. Accounts of such virtuous habits (someone declared, falsely, that every day each tree transpired ten times its own weight!) made the foreign plants seem even more desirable for boggy and fever-ridden areas. In addition to being beautiful and providing useful timber, the trees, experts claimed, had vital healthful and sanitary qualities. Eucalypts were febrifuges, capable of driving away aches, pains, and fevers.

During the 1870s an outpouring of scientific articles, reports, memoirs, and book chapters described how eucalypts restricted and even effectively banished outbreaks of "malaria." That term embraced a range of "agues" and "intermittent fevers," which today we call dengue, yellow fever, typhoid, dysentery, and cholera, in addition to mosquito-transmitted malaria. Until the acceptance of the germ theory of disease, the Western medical establishment believed that the decomposition of organic matter under warm and moist conditions produced toxic gases called malarias, or miasmas. Marshy areas, dense forests in which leaf litter accumulated, and perhaps fields cultivated for the first time, in which upturned soils exposed organic materials, were thought most likely to emit these gases and thereby infect people who worked and lived in these places.

Papers read before learned societies, including the California Academy of Sciences (1872), the French Academy of Sciences (1873), the Royal Botanic Society of London (1874), the Royal Society of Victoria, Australia (1874), the Roman Academy of Medicine (1876), and other medical and pharmacological associations, outlined both the mechanical and chemical processes by which eucalyptus trees acted as febrifuges. Authors also provided medical opinions regarding the efficacy of planting trees, and documented when, how, and where the cultivation of these "fever trees" had induced healthful responses.[31]

Pharmacological, medical, and horticultural periodicals took a keen interest in such opinions, summarized them, and offered details to an attentive public. It appears that the early claims for eucalyptus as a fever-banishing tree can be traced back to Prosper Ramel, an enthusiast who had taken to the genus on a visit to Australia in the mid-1850s; thence to his friend and mentor Ferdinand von Mueller and others in Australia, including Joseph Bosisto, who pioneered the manufacture and export of eucalyptus oil; and also to Sir William MacArthur of Camden Park, Sydney, who attested to the antimiasmic properties of E. globulus.[32]

MacArthur (1800–1882) hailed from a noted family. Born near Sydney, he devoted himself to horticulture, becoming a vintner and exchanger of plants. As commissioner for the Paris Exhibition in 1855, MacArthur assembled Australian woods; as a result of his efforts, he was decorated with France's Legion of Honor and received a knighthood from Britain's Queen Victoria. He firmly believed in the medicinal properties of E. globulus and eagerly distributed seeds, including plantings in the marshes of the Roman Campagna. In 1861 MacArthur reportedly communicated his news about the medicinal qualities of eucalypts to Joseph Decaisne (1807–82), president of the French Academy of Science and director of the Jardin des Plantes, Paris.

Decaisne's friend and collaborator, Charles Naudin (1815–99), who headed the plant laboratory in the Ville Thuret, an adjunct of the national garden in Paris, recalled the importance MacArthur placed on these febrifugal attributes. Naudin cultivated about eighty eucalypt species at Thuret and established close contacts with von Mueller in Australia, and with horticulturist outlets in Algeria. He sponsored the acclimatization of useful plants, such as gum trees, in that French possession.[33]

Von Mueller, who served as government botanist for the state of Victoria, noted that the first "positive experiments" about the therapeutic value of eucalyptus leaves came from Spain, where a "Dr. Tristany" published his medical findings in 1865. Surprised by such positive results, medical practitioners in France, Corsica, Austria, Sicily, Algeria, and Argentina expanded studies of the therapeutic attributes of eucalypts, notably blue gum. They concluded, noted von Mueller, that eucalyptus possessed antifebrile properties and, like pine, would "stay the inflammatory processes in diseases of the respiratory organs," especially in the early stages. It was considered a remedy superior to sea air. In referring to his favorite blue gum, von Mueller concluded with grand rhetoric, "thus is afforded most copiously an oily emanation, befitted to absorb and condense oxygen into ozone, the most powerfully vitalizing, oxydizing and therefore also chemically and therapeutically disinfecting element in nature's whole range over the globe."[34]

With such a battery of expert opinion and adulation, both prestigious and popular journals explored the topic of health and hygiene, and initially accorded eucalypts high marks. A note in the respected British periodical *Nature* (11 June 1874) advised readers that the Italian Government "will gratuitously distribute this year 5,000 plants" of blue gum in a site "infected by malaria." *The Garden,* a popular weekly founded by William Robinson, published ten articles on how the newly termed "fever-gum trees" reduced the prevalence of malaria. Specific mention of improvements referred mainly to the Mediterranean basin, notably North Africa and Iberia, but also included California.

An early column on 12 October 1872 reported, incorrectly it turned out, that hospitals in both Lisbon and Opporto, Portugal, were using bark extract from the blue gum as a substitute for quinine for the treatment for fevers. Another column declared, also inaccurately, that on the island of Mauritius an extract from the same species "was a great service, and a good substitute" for quinine. Such was their fabled success against malaria that in London shops eucalyptus leaves "were sold at sixpence each!"[35]

Subsequent notice of research conducted by a Mr. Broughton, a government "Quinologist" at Madras, India, however, disproved the claims for the quininelike qualities of gum bark. Broughton answered a laudatory article in the British medical journal *The Lancet* (20 April 1872), which claimed that "all parts [of the blue gum] are most valuable as a febrifuge medicine" and that "the leaves, when smoked are most efficacious in allaying pain, calming irritation, and procuring sleep." He challenged the explanation that a bark alkaloid "crystallized like quinine as a sulfate and . . . yielded the ordinary reaction of quinine with chlorine, water, and ammonia," declaring that bark and leaves of blue gum he had worked with carried "neither quinine, quinidine, chinchonidine nor chinchonine . . . in any proportion."[36]

Alkaloids in the bark of cinchona, a genus with approximately thirty species that is native to the temperate zone of the Andes from Colombia into Peru, do interfere with the reproduction of the malarial parasite. Sir Joseph Banks had suggested transferring the quinine-yielding cinchona tree from Peru to Kew and thence into colonial Asia. His scheme came to fruition forty years after his death, when in 1860 four hundred or more seedlings were planted in the Nilgiri Hills close to Ootcamund, where Broughton studied the blue gum that had been introduced almost half a century earlier.[37]

Undaunted by the failure to have his favorites likened to cinchona, Australia-based scientist von Mueller continued to stress the medicinal properties of the genus, and his opinions received wide coverage. The Botanical Society of Edinburgh, for instance, provided a pulpit for von Mueller's recommendation that "fresh branchlets" of mountain ash *(E. regnans),* then named "giant gum," and of other species be placed under the beds of hospitalized patients. They have "the effect [of] being not only antiseptic, but also sedative," insisted von Mueller, "and to some extent hypnotic." He, like others, singled out blue gum, which counteracted "contagia" in the hospital wards. "Eucalyptus leaves generate ozone largely for the purification of the air," insisted von Mueller, "the volatile oil is very antiseptic." In advocating the culture or naturalization of subtropical plants, he also extended the claim for eucalyptus oil as "one of the best for subduing malarian effluvia in fever-regions."[38]

Von Mueller's gaze rested squarely on the Mediterranean basin, especially North Africa, where French settlers and officials, with whom he corresponded, were having good success with blue gum. They planted eucalypts in an attempt to prevent outbreaks of malaria, which reportedly ravaged Algeria between 1867 and 1876. M. Hardy, director of the Jardin d'Essai in Algiers, and the count of Bellerôche sowed the first seeds, which the count obtained from Hardy for his estate of El-Afia in 1862. Almost immediately fevers began to abate. A second fever tree forest established around Lake Fezzara in the early 1870s led to similar results. Ain Mokra, a village on the lake that was used by the French military, grew healthy. The garrison, which had changed every five days due to fevers, reorganized its schedule as soon as sixty thousand or more *E. globulus* set out by the Société Algérienne began to green up the area.[39]

In 1877 British consul-general Playfair of Algiers explained in its "Report on the Febrile Virtues of Eucalyptus" how the genus was a means of reforestation as well as a febrifuge. In the first instance, these fast-growing trees substituted for oaks and other native species that had been felled during settlement a generation or more earlier. The eucalypts grew swiftly and promised a good payoff in timber. In the second instance, "there is no doubt," the consul continued, "of its actions in improving the sanitary condition of unhealthy districts." Playfair was referring specifically to the iron mines of Mokta-el-Hadid, which in the late 1860s had been ornamented with a hundred thousand trees.

Eucalypts worked wonders. Growing trees dried out wet places and wafted into daily breezes "a large quantity of essential oil very similar to turpentine." As a consequence, workers continued to operate the mine during summer months whereas before, due to the unhealthiness of the site, they had taken a train every evening to a barracks 33 km (20.5 mi) away.[40]

Eucalypts pleased the eye in both Algeria and Tunisia, especially in arid and inhospitable places where they took hold and grew rapidly. They formed open-canopied forests, disseminated copious seeds, propagated themselves, and thereby assisted in stemming soil erosion while protecting crops. They are "most favorably naturalized," declared French tree expert Jules Planchon. Estates and gardens could not enclose eucalypts. They spread by "hundreds of thousands, in groves, in avenues, in groups, in isolated stalks, in every section of three provinces [of Algeria]," so successfully that a foreigner "would suppose it to be an indigenous tree," Planchon exclaimed.[41]

"Now most Algerian villages, especially those in malarial regions," declared resident Edward Pepper, M.D., "have more or less extensive groves or avenues of eucalypti." Pepper explained how balsamic odors worked. The plants exhaled an essential oil, which oxidized the air into ozone. Twigs and leaves supplied a camphorlike eucalyptol, "most serviceable as a febrifuge, tonic stimulant, aseptic and antiseptic."[42]

Similar confirmation of eucalyptus trees reducing the incidence and severity of malaria in both North Africa and Corsica came from other credible sources. At a reunion at the Sorbonne in Paris, Dr. de Pietra Santra, of the Climatological Society of Algiers, spelled out results from a survey of fifty Algerian sites with a combined total of a million trees. The "sanitary effect" was unquestionable, he concluded, intermittent fevers had abated, and large tracts of previously useless land had been reclaimed by new forests. The same was happening in Corsica, where six hundred thousand fever trees were growing.[43]

Notices from South Africa and Cuba, where eucalypts had thwarted "paludal and telluric diseases" (that is, those caused by marshes and soils), spelled out similar advantages. Wherever eucalypts grew, malaria was forced into retreat and submission. There was even an incorrect belief (attributed to Prosper Ramel) that eucalyptus subdued malaria in its native home of Australia.[44]

Accolades continued to accrue from illustrious gatherings. In 1874 a professor of botany and dean of the medical faculty at King's College, London, R. Bentley, lectured to the Royal Botanic Society of London about eucalypts and their medicinal properties; his address appeared verbatim in *The Garden* (2 May 1874). Bentley, an authority on his subject, quoted numerous examples of the tree's "suction-powers" through its "extensive piping" underground. The Australian trees acted as huge sponges in draining febrile areas, while at the same time their leaves released "odorous antiseptic emanations" that neutralized marsh miasmas. The professor concluded that one could say "without exaggeration" that "this tree does possess a most beneficial effect in neutralising and improving the malarious influence of marshy districts."[45]

Bentley also drew attention to experiments that proved the solidity, hardness, and durability of eucalyptus wood and noted that potash fertilizer derived from gum trees proved superior to any from elms and maples (in the United States). Additionally, oils in amounts of a thousand pounds per month were leaving Melbourne, Australia, for use overseas in medicines and perfumes. "The genus," he exclaimed, "must be regarded as one of the most important to man in the vegetable kingdom."

Another article in *The Garden* of 30 October 1875, "The Gum Tree in One of Its New Homes," explained that the fashion of planting eucalypts in California was in large part due to "its supposed influence in absorbing the malaria." Other factors attracted investors, but the link to health was crucial. As stories like these appeared in the popular press, sales of "fever gums" proceeded briskly in the United Kingdom and in the United States, and enthusiasts continued to wrangle about which eucalypts were the most hardy and how to obtain identifiable seeds.[46]

One correspondent to *The Garden* notified colleagues that he intended to send seeds to the River Gambia, West Africa, to fight malaria. Another remarked on how lines of 40 ft (12.2 m) high *E. globulus,* flowering on the French Riviera around San Remo, Nice, and Cannes, exhaled "a rich resinous aroma" that reduced fevers. Everyone, it seemed, recognized and benefited from these foreign giants that grew best in the warmest parts of Provence. Enthusiasts planted them in about 1858. The Huber bothers grew them in Hyères, as did Gustave Thuret in Antibes. In response to this kind of support, the French Société d'Acclimatation offered a prize of 500 francs for "a practical and theoretical guide to the cultivation of the Eucalyptus." To compete, growers had to have been experimenting with the plant genus for at least five years.[47]

Reassessment in the British Isles

The hyperbole used in promoting the importance of eucalypts to forestry and public health began to unravel in the 1880s. In Great Britain and Ireland doubts centered on growth and longevity in cold, wet, and marshy areas. Growers, it seemed, had a hard time keeping the trees alive.[48]

Questions began to arise as to whether any part of Britain and Ireland could benefit from *Eucalyptus*. A query in *The Garden* about what species was most desirable elicited the reply that it is "not very hardy, nor beautiful, therefore, [it] really shouldn't be as popular as it is." In the late 1870s the earl of Annesley planted nineteen eucalypt varieties in his garden in Castlewellan, County Down, Ireland. Twenty-four years later he concluded, "I cannot do much with the Eucalyptus." Blowdowns and spindly growth had destroyed them. English and Irish eucalypts failed to attain the stature or grace of similar trees in California or southern France. By the early 1900s plant authorities concluded with customary understatement that the blue gum is "perhaps not absolutely hardy in any part of Britain or Ireland."[49]

A notice entitled "Eucalypts" that appeared in *The Garden* (4 June 1881) pierced the commercial bubble. It scolded the London *Times* for printing inflated and preposterous claims about the benefits of these Australian trees. It asserted that "the value of the tree was long ago settled in our gardens, even before our recent hard winters began to make their mark." Eucalypts survived in favored spots for a few years and then died from frost injury. *The Garden* accused the *Times* of unethically publishing faulty advice that misled readers into purchasing thousands of trees every year, only to see most of them perish.[50]

The same situation existed on the European continent. In France, M. Tourasse of Pau had tested fifty-five species and shown that only one *(E. coriacea)* could withstand hard frost. It was certain that the genus was not able to be acclimated in northern Europe, although nurseries, it seems, were loath to admit it. For example, French nurseryman Henri Leveque de Vilmorin's 1882 catalog listed more than eighty eucalypts in which the seed expert had become interested while in Algeria, even though experience demonstrated that no more than four or five were in general cultivation, and then solely for the conservatory. Thus, irrespective of the "sanitary" attributes of gum trees or their manifold uses for timber, most, if not all, were ill suited to survive a cold European winter.[51]

By 1890 disappointment reigned throughout the British Isles. In Scotland, for example, J. Muir from Margam concluded that "all are agreed that this tree gives no promise of becoming a permanent figure in British gardens." Back from a recent visit to Lord Dalrymple's estate, Muir reported that several large trees grew at Loch Inch Castle, seat of the lord's father, the earl of Stair, in Wigtownshire, but he knew of no others. Other trees, however, did exist in Scotland. And against all odds, trials were still being conducted with a dozen or so, for example, on the Isle of Arran, in Argyllshire, and near Greenock.

At least eight specimens of the cider gum *(E. gunnii)*, so called for the beverage made from its sap, were alive, and some thirty-five additional species were being raised at 57°N latitude by Robert Birkbeck, banker, respected botanist, and proprietor of Kinloch Hourn on Scotland's west coast. Birkbeck concluded that *E. gunnii, E. coccifera, E. cordata,* and *E. urnigera* were "quite hardy," although not as tough as *E. vernicosa.* Blue gums could barely withstand fifteen degrees of frost. Birkbeck's optimism, however, was premature. The cold winter of 1894/5 destroyed most of his trees, even in his greenhouse (although at least four specimens survived into the 1970s when author Robert Zacharin observed them).

Birkbeck raised his plants under glass, then repotted them to keep them from becoming rootbound. When the eucalypts were 3 ft (1 m) high, he staked them out with loose fastenings to prevent them from being "easily blown down." Other growers in Scotland at that time included the Established and Free Church ministers at Roseneath, Dumbartonshire, and James Paterson, commissioner in Arran to the duke of Hamilton.[52]

Promoters in France

In *Emigrant Eucalypts,* Robert Fyfe Zacharin identifies France as "a key distributing point for secondary transfer." Unlike the United Kingdom, which is warmed by the Gulf Stream, average winter temperatures on the European continent normally proved too harsh for most species, including the popular *E. globulus,* upon which early collectors and botanists heaped special hopes and praise. Nevertheless, many promoters recommended the genus for botanical gardens and private estates, especially around the Mediterranean Sea. Odd-looking or rare species drew onlookers. The origins of these plants in such a strange and mysterious quarter of the Southern Hemisphere added to their cachet. Significantly, in public exhibits and displays, eucalypts symbolized the political and environmental hegemony of governments whose reach could span thousands of miles. Collectors and their wealthy clients made important shipments of Australian gum seeds to nations in southern Europe and North Africa. French botanists led the charge.[53]

Botanist Jacques-Julien Houttou de La Billardière (1755–1834), who first described the blue gum *(E. globulus)* from a specimen he collected in Tasmania in 1791, was one of this group. Born in Alençon, and turning to botany rather than medicine at the university of Montpellier, La Billardière realized his dream of exploring Australia when he joined an expedition to search for the explorer Jean François de Galaup, count of La Pérouse. This French aristocrat had sailed from Brest in August 1785 bound for the Pacific Ocean under instructions from Louis XVI to verify new discoveries. Authorities had received last word from him at his six-week anchorage in Botany Bay in February 1788, but as time passed without further news, in 1791 the French National Assembly decided to organize a rescue mission. (Forty years later, it was learned that both expedition vessels had been wrecked in the Santa Cruz group, probably in that year.)

Tracking La Pérouse to Australia, La Billardière landed in Tasmania in May 1792, where he came upon a beautiful, smooth-barked new eucalypt whose fruit "resembles a coat button in shape." The crew felled one in order to obtain its blossoms and noted a copious flow of sap from the center of the lower trunk. La Billardière judged the blue gum one of the tallest trees in nature, suited for naval construction, specifically masts. He noted that it was a beautiful tree with a smooth bark. "The branches bend a little as they rise and are garnished at their extremities with alternate leaves, slightly curved, and about seven inches in length, and nearly two in width. The flowers are solitary, and grow out of the axils of the leaves. The bark, leaves and fruit are aromatic, and might be employed for economical uses in place of those which the Moluccas have hitherto exclusively furnished us."[54]

Early the following decade, two additional French botanists joined a cartographic survey along the Australian coast and collected plants for Malmaison, the famous garden belonging to Empress Josephine. In their survey of native plants, they mentioned two eucalypts, red mahogany *(E. resinifera)*, described as a remedy for dysentery, and swamp mahogany *(E. robusta)*, a potentially valuable forest tree. It is unclear how many eucalypt seeds actually ended up at Malmaison or its equally renowned contemporary enclosure, the Jardin des Plantes in Paris (originally the Jardin du Roi, founded in 1635 by Louis XIII's physician Guy de la Brosse). When Aimé Bonpland, who was appointed to conserve the garden plants at Malmaison, made his *Description des plantes rares* (1813), he mentioned only one eucalyptus, but was "aware of 24 different species" described by earlier authors.[55]

Prosper Ramel, member of France's Zoological Society for Acclimatization of plants and animals, visited Australia in 1854 and formed a lasting friendship with "that indefatigable botanist of the Antipodes," Ferdinand von Mueller. He was greatly impressed by the appearance and vigor of *E. globulus* shown him by von Mueller's superior John Dallachy, superintendent of Melbourne's expanding Botanical Garden. Von Mueller, botanist for the state of Victoria and the garden's increasingly famous collector, was on an expedition when Ramel visited. On his return von Mueller struck up a friendship and urged the French trader to make blue gum and similar Australian hardwoods famous throughout Europe.

Ramel returned to France in 1857 infatuated with the grace and beauty of eucalypts, especially *E. globulus.* He delivered seeds to the National Museum, and to the Paris-based Society for the Acclimatization of Plants and Animals of which he was an active member. Traveling widely throughout the Mediterranean, Ramel campaigned to spread the reputation and advertise the utility of eucalyptus. He even encouraged cigarettes to be rolled from chopped-up leaves. Later, American eucalyptus benefactor Abbot Kinney referred to him as "the apostle of the eucalyptus," while von Mueller was "its prophet."[56]

In France, eucalypts did best in Provence, along the Italian border, and in Corsica. Take-off had occurred in the 1860s and expanded into hundreds of thousands of trees. Horticultural expert Naudin listed thirty species in Provence, and referring to Spain, Italy, and Sicily, he reported "immense plantations" already given over to them. Most people loved the color and shape of *E. globulus,* which abounded in the area around Toulon. Testimonials from medical practitioners in Toulon, Cannes, Montpellier, and Paris about the medicinal value of its leaves established in Naudin's mind that "*E. globulus* is an invaluable tree, the reputation it has for improving unhealthy regions and preventing marsh fevers, if a little exaggerated, is not completely without merit."

The scientist-cultivator praised "the famous and generous" von Mueller for spreading "huge quantities of *Eucalyptus* seed" both directly and through intermediaries, such as Ramel.[57]

Before Ramel's encounter with von Mueller, individual eucalypts sprouted in southern Europe. A blue gum specimen was reportedly growing in the hills above Naples, Italy, on the estate of the count of Camaldoli as early as the 1820s, when the estate's head gardener, Frederick Dehnhardt, named it *E. gigantea.* Two additional species flourished there. One of them Dehnhardt named *E. ambigua,* which von Mueller identified later as peppermint gum *(E. amygdalina).* The other Dehnhardt called *E. camaldulensis*; von Mueller identified it as river red gum *(E. rostrata),* appropriately named "beaked" after its elongated flower bud. The botanist obtained all three species to include in his major book on eucalypts, from Baron Vicenzo Cesati, director of the Naples Botanic Gardens, who had received seeds from the Camaldoli estate. Dehnhardt's scientific name *E. camaldulensis* has taken precedence for what is the most widespread of all the eucalypts in Australia and perhaps in the world. It is common along the Murray-Darling River basin of southeast Australia, and occurs in other riparian areas where it stands as a medium-sized to tall tree.[58]

One must credit the diffusion of much of the genus in western Europe to Ramel, who pushed enthusiastically and tenaciously for the expansion of blue gums, in particular on a "forestal scale." The whole enterprise, however, was the brainchild of his friend and confidant Ferdinand von Mueller, who took a close professional interest in the genus, discovering, naming, classifying, and assessing the economic potential of approximately one hundred twenty species over a forty-year-long professional career.[59]

Beyond Australia

Ferdinand von Mueller

Every country has its own landscape which deposits itself in layers on the consciousness of its citizens . . . so many eucalypts have been exported to different countries in the world, where they've grown into sturdy see-through trees and infected the purity of these landscapes. Summer views of Italy, Portugal, Northern India, California, to take obvious examples, can appear at first glance as classic Australian landscapes—until the eucalypts begin to look slightly out of place, like giraffes in Scotland or Tasmania.

—MURRAY BAIL,
Eucalyptus: A Novel, 1998

A steady drumbeat of publications about the taxonomy, natural history, cultivation, and management of eucalypts, plus expert recognition that the genus was able to adapt to a range of physical conditions, enhanced prospects for cultivation outside Australia. As we have seen, the "pull" was from European-based botanists, collectors, and wealthy landowners, who delighted in new and potentially useful plants and readily exchanged details about cultivation and care. The "push" involved a handful of Australia-based scientists, who directed the plant genus toward areas in need of fuelwood, timber, wood-based products, and reforestation. There was an overarching confidence on the part of both in their abilities to reorganize and improve the natural world. Information about novel plants and animals and methods for transporting them, and assessments of potential contributions to a nation's capital and prestige stoked this optimism. Nature awaited manipulation, and colonial areas were sources for discovery, collection, and extraction. They were also testing grounds for familiar plants and animals — domesticated or otherwise — to furnish the material necessities and comforts for pioneers and settlers. One of the most energetic and influential of the Australia-based boosters was Ferdinand Jakob Heinrich von Mueller (1825–96), government botanist, explorer, author, plant collector, taxonomist, and promoter, a man with a penchant for eucalyptus.

This humbly born German scientist devoted a long and successful career to the study of Australia's flora. Von Mueller ransacked his adopted home state of Victoria for "green gold," collecting, identifying, and naming thousands of new plants from around Australia. He sought also to "improve" native flora and fauna by importing useful plants and animals. In 1857, as government botanist of Victoria and newly appointed director of Melbourne's Zoological and Acclimatization Society, von Mueller encouraged the importation of useful plants and animals from Europe, North America, and elsewhere. In the following years, when he also served as director of the Royal Botanic Gardens, he despatched from his Melbourne base live plants and packets of seeds of the Monterey pine *(Pinus radiata)* to growers in the state of Victoria.[1]

This conifer is naturally restricted to three small coastal refugia in central California, the largest of which is on the Monterey Peninsula and the two others on Mexican islands off the coast of Baja California. Over the past century or so, through the efforts of botanists like Mueller, the Monterey pine has risen in economic importance to become, like eucalyptus, one of the most remarkable successes in plantation forestry. Foresters claim that it is the most widely planted tree in the world. It flourishes as a staple softwood in the Southern Hemisphere, mainly in New Zealand, Chile, Australia, and South Africa. Like eucalyptus, *Pinus radiata* has been a component of wood-growing schemes in Spain, Portugal, and Kenya.[2]

Von Mueller believed that alien trees, together with mammals and birds, would benefit European settlers. He applauded introductions of familiar passerine birds from England, advocated the introduction of the alpaca and the cashmere goat to Australia's "grassless forest-ranges," and argued that "desert game" mammals from South Africa would thrive on the nation's "inland solitudes." He expressed an evangelical duty to move humanity (and his own fame and reputation) forward according to God's law. Scientific knowledge drove progress, and armed with methodology, discipline, and pietistic beliefs implanted by his north German Protestant upbringing, von Mueller accorded to himself and fellow humans the role of architects of a Christian "calling" to be modifiers and improvers of both society and nature.[3]

A Passion for Eucalypts

An accomplished plant collector and botanist, who submitted a thesis on the common shepherd's purse *(Capsella bursa-pastoris)* for his doctorate at Kiel University at the tender age of twenty-one, Ferdinand Müller, as he was originally named, also received a degree in pharmacy. His mother arranged for him to apprentice with pharmacist A. G. Becker in Husum (from which he attended Kiel). In 1847 the young man reportedly decided to emigrate to Australia after discussions with Johann Ludwig Preiss, who had organized an expedition to the Swan River Colony (Western Australia) a decade earlier in order to collect plants and animals. He had secured investors, set sail, and landed in 1838. For the next four years Preiss traveled, collecting more than a hundred thousand specimens, but was continually dogged by financial difficulties. In Perth, he offered his booty to Governor Hutt for 3,000 pounds sterling. Rebuffed by the governor, Preiss still managed to sell his collection, and he finally returned to Germany. Whatever his experiences in Australia, his retelling them to Müller piqued the young scientist's spirit of adventure and curiosity. Müller was especially interested in descriptions of Australia's plants, as well as in its salubrious climate.[4]

Müller's father, a former Prussian army officer and customs agent in the port of Rostock where Ferdinand was born, died of tuberculosis in 1835. His mother Louise met the same sad fate in 1840, as did his older sister Iwanne in 1843. These tragedies helped spur his decision to leave for Australia with his surviving sisters Bertha and Clara. Asthmatic and bronchial problems also vexed him. Ferdinand and the two young women left Bremen in July 1847 aboard the *Hermann von Beckerath,* bound for Adelaide. They anchored for a week in Rio de Janeiro, where the twenty-two-year-old scientist scrambled after plants. On 15 December, as the ship docked in Adelaide, he took specimens of seaweed from the harbor. The trio took up lodgings in the rough port. Ferdinand found a job for 15 shillings per week in the chemist's shop on Rundle Street belonging to Moritz Heuzenroeder, whose brother had made the voyage with them. There, he supported himself and his sisters, and indulged his passion for plant hunting whenever there was an idle moment. Frail-looking Bertha, who was twenty-one, and the more robust fourteen-year-old Clara helped press specimens and organized their elder brother's collection. Except for a brief, unsuccessful stint at farming, Ferdinand worked as a pharmacist in Adelaide for the following five years. Then he decided to strike for nearby goldfields in the state of Victoria, where he intended to open his own pharmacy.[5]

At that time, in Melbourne, the capital of Victoria, Lieutenant-Governor Charles Joseph La Trobe was searching for a scientist to take charge of the state's new and expanding plant collection. La Trobe liked the qualifications, plant knowledge, and enthusiasm of his new young acquaintance, and in January 1853 he appointed Müller to be the first government botanist, with an annual salary of 400 pounds sterling. There is a debate as to whether Sir William Hooker, protégé of Sir Joseph Banks and director of London's Royal Botanic Gardens at Kew, assisted in von Mueller's appointment. Reportedly, von Mueller's paper on the flora of South Australia, translated into English and read before London's Linnean Society, may have led Hooker to recommend him to La Trobe. Recent research has found no documentary evidence that Hooker knew about von Mueller before his appointment by La Trobe. The governor appears to have had freedom in such matters. Probably Hooker learned of the young scientist after von Mueller introduced himself in a letter he sent to Kew shortly after his appointment. As government botanist, von Mueller began to correspond on scientific matters with Hooker and received encouragement and support from his mentor, until Hooker's death in 1865. By the time he took up his new post, Müller had anglicized his name to Mueller, and had become a naturalized British subject.[6]

Mueller, as he now styled himself, worked out of Melbourne's Botanic Garden, serving as its director from 1857 to 1873. The Botanic Garden, which had been founded a year before the Müller family had landed in South Australia, was expanding rapidly and already displayed a collection of a thousand native and five thousand foreign plants in an 83 acre (33.6 ha) enclosure. Governor La Trobe took a keen interest in the specimens growing there.[7]

Mueller's passion for botanizing found an instant outlet in Australia. Everything was new. Plant collecting, description, identification, systematics, and iconography were all in their infancy. The Melbourne appointment enabled the energetic twenty-eight-year-old to throw himself into botany full time.

Mueller initially concentrated on collecting unknown plants. He joined excursions and expeditions to discover, name, and describe new species. His biographer Alec Chisholm calculated that Mueller "covered on foot and on horseback at least 15,000 miles of the Australian landscape" in his first years in government service, compiling at least forty-five thousand plant specimens. He slogged through the rough terrain in the Northern Territory as official botanist with the Gregory Expedition in 1855–56, as well as in his adopted state of Victoria. Mueller dedicated himself to plant collection and description for more than forty years. In the words of a recent author evaluating the German influence on botany in Australia, "it is scarcely an exaggeration to say that he covered Victoria on his hands and knees. Few species went unnoticed by this ardent collector and observant botanist."[8]

Mueller rapidly achieved international recognition for pioneering plant science in Australia. He authored or collaborated on fifteen major publications in botany, including *Eucalyptographia: A Descriptive Atlas of the Eucalypts of Australia* (published in ten parts between 1879 and 1884). In 1867 the king of Württemberg conferred on him the title of baron for his scientific discoveries (thereafter he signed himself "F. von Mueller," using the particle of nobility). A little over a decade later, Britain's Queen Victoria recognized his contribution to science and services to crown and country by bestowing upon him the title of Knight Commander of St. Michael and St. George. Altogether, von Mueller named or reclassified 4,700 species, penned 1,500 publications in 110 different periodicals, and carried on an unremitting stream of correspondence (averaging 3,000 or more letters per year) with colleagues and learned societies all over the world. He increased the botanic garden's herbarium holdings tenfold from forty-five thousand specimens in 1857 to five hundred thousand by 1888.[9]

Much of von Mueller's work centered on the natural history and benefits of growing eucalypts. His favorite species was the blue gum, which he termed the "Prince of Eucalypts." This energetic and visionary scientist paid attention to both the aesthetic and economic value of these plants, including their oils, and consulted with colleagues in Australia and around the world who had expertise and experience in these topics. In addressing the commercial importance of eucalyptus oil, for example, von Mueller cooperated with Joseph Bosisto, who built the first factory for large-scale extraction of the oil near Western Port in 1862.[10]

Bosisto (1827–98), an Englishman and fellow pharmacist, who studied in the School of Medicine at Leeds, emigrated to Victoria in 1848. Most probably Bosisto encountered von Mueller through pharmacy. Bosisto patented a process for extracting the "greatest amount of the purest essential oil" from eucalyptus foliage, "with the least consumption of fuel and application of labor," noted his friend. The two eucalyptologists collaborated on a project to distill eucalyptus oil for the state of Victoria's Industrial Exhibition in 1854. One effort led to another, and Bosisto, with advice from von Mueller, experimented with about thirty oils, which he displayed in another exhibition a few years later. He studied their properties as solvents, preservatives, fuels, and health remedies. Eucalyptus oil and soaps made scented toiletries, and this Yorkshire-educated entrepreneur sought out markets in England, Germany, and the United States for his oil-of-eucalypt products. He showed off his intentions in fairs and expositions in cities such as Philadelphia, Vienna, and Paris, bringing to international attention the "eucalyptus-medicines." By the early 1870s, Bosisto's factory was producing 700 pounds (317.5 kg) of oil every month.[11]

One reason Bosisto, von Mueller, and others found eucalypts so interesting and attractive was expressed in a public lecture in London's Conference Hall, South Kensington, delivered by Bosisto on the occasion of the Colonial and Indian Exhibition in 1886. Bosisto, who was then a member for Richmond in the legislative assembly of Victoria, had lived in Australia for almost forty years. He admitted in his lecture that the countryside of his adopted home lacked the lush greenery of England, but maintained that it had its own interesting and variegated flora, of which eucalypts were the most significant elements. Eucalypts typified the character and identity of Australia, dispelling the monotony of its scenery through "interesting variations in the leaf formation, in the colour both of leaf and flower, and in the appearance of the tree bark, and the shape and varied stature of the trees." Bosisto reiterated the opinions of artists, notably watercolorists, who, for more than fifty years, had celebrated eucalypts as "true Australians." Eucalypts expressed the resilience of the outback or "bush," provided useful implements for colonists, and were becoming a pastoral backdrop for settlement and ranching.[12]

Popular watercolorist Hans Heysen, for example, depicted Australia's landscapes through the textures and shapes of native gum trees, noting how light picked out various hues and textures of trunks and branches. He paid special attention to the twisted, gnarled, and pitted trunks of mammoth specimens, under whose foliage livestock sought shade.

Like Heysen, Bosisto saw *Eucalyptus* as "a firm friend of man and beast," a helpmate for European immigration. He exclaimed that "many a meal of damper and mutton, and many a smoke pipe of peace have been partaken inside their cavernous recesses." Like his German friend, Bosisto singled out the blue gum for recognition. It was stronger than oak or teak, he declared, and a whole set of products—from bridges, docks, and ships to houses, wagons, and fences—could be fashioned from the wood of this gum species.[13]

Von Mueller's enthusiasm for eucalypts and his willingness to recruit and cooperate with people like Bosisto led him into a variety of timber-related fields. The wood from his favorite blue gum was pale in color, but hard, heavy, strong, and durable, he declared. "In house-building it is one of our best timbers," he noted, referring to joists, studs, rafters, and heavy scantlings sawn by Australian carpenters. Lloyds of Australia gave blue gum a second-class rating for colonial timber, he reported (three other eucalypts received the highest rating of first class), yet ships with decks, beams, hooks, sternposts, transoms, and deck-waterways built from blue gum could expect those pieces to last at least ten years in tip-top condition. Builders crafted the same wood extensively for carriages and agricultural implements. "It is further used in telegraph-poles and for planking for bridges and jetties," declared von Mueller. Railway sleepers and fencing rounded out a long list of timber products to be made from this gum.[14]

Such reports, bolstered with information about propagation and dissemination, increased the reputation of the plant genus overall, and drew interest and attention to a range of species. Von Mueller and colleagues continued to trumpet the virtues of each one. Southern or bastard mahogany *(E. botryoides)* has "timber usually sound to the center, adapted for water-works, wagons, particularly felloes [the outer rim of a wheel supported by the spokes]," von Mueller stated. Karri *(E. diversicolor)* is good for rafters, spokes, and fence rails; *E. cornuta* is valuable for "various artisan's work, and is preferred for the strongest shafts and frames of carts." And the reddish, hard, heavy, yet elastic wood of the narrow-leafed red ironbark *(E. crebra)* is "much used" in bridges and railroad ties.[15]

This naturalized Australian knew them all. He arranged 117 species systematically, admitting that there were twenty to thirty "kinds" yet to be illustrated, and showed his devotion for the genus by seizing every opportunity to promote its physical, medicinal, and aesthetic attributes. His knowledge and experience with the blue gum, his generosity in supplying seeds to almost everyone who requested them, including details about how to grow them and get the best results, plus an unflagging ebullience and energy, proved crucial for the successes that followed. Characterized correctly as "the most renowned student of the great Australian genus Eucalyptus," von Mueller gave unstinting praise to the tree.[16]

In the introduction to his *Eucalytographia* (published from materials collected after 1847), von Mueller intended to demonstrate the importance of eucalypts, "whether viewed in their often unparalleled celerity of growth among hardwood-trees, or estimated in their manifold applicabilities to the purposes of industrial life, or contemplated as representing among them in all-overtowering height the loftiest trees in her Majesty's dominions." In his extensive treatment of the plant genus, the famous botanist prophesied that the trees were going to play "for all times to come" a prominent role in the "sylvan culture" of vast land tracts, a role that would include supplies of hardwood, "sanitary measures," and "beneficent climatic changes." It was to wood products and health that he directed the bulk of his statements about applied forestry.[17]

Diffusion

Von Mueller corresponded with and supplied seeds to eminent foresters in North America and Europe. One of them in the United States was federal horticulturist William Saunders, cofounder of the National Grange. The Scottish-born Saunders grew some blue gums from seeds that von Mueller sent him, and in turn collected seeds from trees he had grown. These he distributed for the U.S. Department of Agriculture in response to a growing public interest in eucalypts.[18]

The Melbourne-based scientist mailed seeds to friends and acquaintances. Prosper Ramel used them for afforestation projects in Algeria and Tunisia. When Sir William Hooker complained how sparse trees were around the holy city of Jerusalem, his protégé sent him seeds of eucalyptus trees. Von Mueller was in touch with experts in France, India, South Africa, Latin America, and elsewhere. The extent of his network of connections and his energy is exemplified by the fact that in 1861 alone, von Mueller arranged to send 22 cases of plants overseas, containing 31,000 live plants, 36,400 cuttings, and 51,290 packets of seeds from Melbourne's Botanic Garden.[19]

The most famous story that links von Mueller to the spread of his favorite plants comes from Italy. In 1869 Archbishop J. A. Gould of Melbourne attended the Vatican General Council in Rome. He arrived with eucalypt seeds (probably from the river red gum *[E. camaldulensis]*, or perhaps blue gum) supplied by von Mueller, and presented them to a community of French Trappist monks in the monastery of Tre Fontane. The monastery was reputed to be "a fever-stricken locality." French contemplative priests had recently taken possession of this sacred site, which reportedly marked the spot where Saint Paul was beheaded. Wishing to honor this tradition, the foreign-born clerics soon discovered why no Italian order was willing to settle there. It was, in their words, the most unhealthy spot in the entire Roman Campagna.

Despite these conditions, the Trappists refurbished buildings, drained surrounding lands, and planted crops. Several brothers succumbed to agues, others gave up and returned to France, while the remainder gave serious thought to abandoning the project. Eucalyptus trees, with their growing reputation for banishing swamp fevers, offered hope. The countess of Diesbach, one of the tree's admirers, supplied instructions for cultivation, and the archbishop of Melbourne furnished the seeds.

Initially, due to neglect and the monks' inexperience, most of the seedlings died, as did a second batch of sixty supplied in 1870 by a French lady, Madame Cutheu. However, within a decade, fresh eucalyptus plants had begun to work wonders—drying out the swampy land and reducing the incidence of malaria. Spurred on by crop innovator Abbot Timothy of Grande Trappe, Brother Gildas, who was charged with tree planting and tending, reported excellent results. This occurred despite predictions from some locals that eucalypts were ill suited to the region.

According to Archbishop Gould, eucalypts acted "most wholesomely on the poisonous air of that part of Campagna," and reduced the marshy character of its environs. On a later voyage to Italy, the prelate revisited Tre Fontane and was astonished by the changes. He declared that trees from Australia had done what several Roman emperors long before the Christian era, even Caesar himself, had sought to do but had failed to accomplish—namely, banish fevers from the Eternal City.[20]

Another version of this story was not quite as complimentary to the trees. In his report, Brother Gildas was guarded about whether eucalypts had actually banished malaria around Tre Fontane. Improvements, he noted, were associated more with drainage and cropping than with tree planting. Taken together, these changes had made malaria less prevalent. Nevertheless, Gildas admitted that people visited the Trappist monastery to seek his counsel about cultivating "miracle" eucalyptus, and he had no reservations about recommending them. One famous pilgrim was Prosper Ramel, renowned French propagator and ardent grower. Another was Prince Troubetzkoy, who grew eucalypts on his estate on Lago di Maggiore. Such well-known visitors helped establish and spread the good name of eucalypts. Cultivation expanded around the hallowed site as thousands of additional trees went in, adding to the fame of Tre Fontane. In response, the Italian government endorsed projects for planting eucalypts in other areas.[21]

Debate Intensifies

Despite supposed successes like Tre Fontane, to which von Mueller referred in some of his commentaries and about which he had no misgivings, disagreement about the healthfulness of eucalyptus continued. In 1894 the U.S. State Department issued an instruction to consuls requesting information about the value and uses of eucalyptus. Wallace S. Jones, consul-general in Rome, responded. He dismissed the claim that eucalypts imparted a "balsamic atmosphere" healthful to humans. The Trappist experiment on the outskirts of Rome had failed to quash malaria, he argued, as was proven by outbreaks that occurred into the 1880s—a good fifteen years after initial plantings. A government prison built close to the monastery in 1880 had run into difficulties after guards and inmates had contracted the disease. The American official cited a Roman physician attached to the prison as declaring that "no beneficial result against malaria has been derived from the planting of the *Eucalyptus*." Jones concluded that "all this enthusiasm for the *Eucalyptus* tree is in no wise justified." The tree was "capricious," it often died after frosts, was split by moderate winds, showed uneven spurts of growth, and usually proved costly to cultivate. Its improvements to the air were no more advantageous, he concluded, than those of the much hardier, adaptable elms, pines, or mulberries.[22]

Jones's opinions were bolstered by additional other reports debunking eucalypts. One came from a medical practitioner who in a letter to the *British Medical Journal* in 1884 claimed that drainage, not eucalypts, suppressed malaria. Another referred to Rome-based professor Tommasi-Crudeli's opinion, delivered to the International Medical Congress also in 1884, that eucalypts had been ineffective in controlling a severe outbreak of malaria.

However, American pharmacist and public hygienist Charles Kingzett, who knew about negative reports, favored the Australian tree. He concluded that "because malaria is an infectious disease, the poison of which may be conveyed in the air and in water . . . it would be necessary to not only have plantations of eucalypts covering the whole area under observation, but also shut off contamination from exterior sources." His conclusion (repeated as late as 1907 in the fifth edition of his book *Nature's Hygiene*) was that "the eucalyptus certainly gives some protection, even if immunity cannot be claimed for it."[23]

Although apocryphal stories about the vaunted benefits of growing gum trees usually lead back to von Mueller and his circle of like-minded enthusiasts, von Mueller himself was careful about stating overtly that the genus was a cure for malaria. He let others, such as Archbishop Gould, make these forthright claims. In *Eucalyptographia,* von Mueller published the names of ten physicians in France and around the Mediterranean who had conducted clinical experiments that suggested that the tree was indeed a febrifuge. He noted that early medical tests for its effectiveness against malaria dated back to 1865, but concluded that there was "no accurate pathologic data" on the effects of the "exhalation of Eucalyptus-forests on phthisic [consumptive] patients." He suspected, however, that odors from live trees did curtail or arrest diseases of the bronchia and lungs.

Others claimed outright that eucalyptus cured malaria or so-called intermittent fevers. For example, Ramel in 1872 quoted a physician, Dr. Adolph Brunel of Toulon, about the value of *E. globulus* in suppressing agues. Publications out of Spain and reports from places as far-flung as Latin America and West Africa attested to the merits of this species as a febrifuge. From medical results in Paraguay it was clear, according to Brunel, that aromatic emanations from blue gum "neutralized swamp effluvia." The physician believed that the trees reduced the extent of swamps through their energetic absorption of water and that patients with bronchial and respiratory afflictions benefited from medications containing eucalyptus oils. These sorts of claims continued to be repeated and elaborated upon throughout the final quarter of the nineteenth century.[24]

Von Mueller was far more emphatic about the environmental benefits of growing his favorites. He stated that woods or forests of these trees acted like salves, wafting into the atmosphere oils that were "the most powerfully vitalizing, oxidizing and therefore also chemically and therapeutically disinfecting element in nature's whole range over the globe." Climate could be improved and ameliorated, he noted, by planting barren lands with the woody plants.[25]

Marketing

Despite disagreements over its medical value, the market for eucalyptus continued to expand, due in part to the special qualities of eucalyptus seeds. In the early 1870s an industry had grown up based on commerce in the seeds of acacia, casuarina, and, most important, eucalyptus. "What gives to our own export trade of forest seeds such significance," von Mueller declared, "is the fact that we offer thereby the means of raising woods with far more celerity and ease than would be possible through dissemination of trees from any other part of the globe." He stressed that for eucalyptus seeds "scarcely any care [is] requisite in nursery works." Within a year or even a few months, the young plant was ready for transplantation, and the "minuteness of most kinds" of seeds from which they sprang and their natural dryness made them easy to pack and ship. Apparently, the pests and diseases were left in Australia.

With an eye on the commercial potential of raising his preferred trees, von Mueller instructed an assistant to count the number of seeds in one ounce and concluded that blue gum furnished 10,112 seeds per ounce (357 per gram), peppermint tallied 17,600 (621), and swamp gum a record 23,264 (821) seed-grains. Calculations suggested that one pound of blue gum seed could result in 161,792 trees; if merely one-quarter of them germinated, there would be enough seedlings to cover 404 acres at 100 trees to an acre. "It seems marvelous," he concluded, "that trees of such colossal dimensions, counting among the most gigantic of the globe, should arise from a seed-grain so extremely minute." Traders had begun to export seeds profitably, and von Mueller estimated that the value of shipments abroad (he listed London, Paris, Calcutta, San Francisco, Valparaiso, and Buenos Aires as clearinghouses) were worth thousands of pounds sterling.[26]

This seed trade continued to grow. Faced with problems of cutover land, rapid deforestation, and increasing needs for wood for both fuel and timber, government officials in many countries began to organize schemes for reforestation by utilizing adaptable, reliable, and useful tree species. Eucalypts fitted this profile. Officials were struck by the fervor of industrial Victorians such as von Mueller, who pioneered botanical research, categorized plant benefits and uses, and argued how easy it was to grow, establish, tend, and expand the areal extent of eucalypts. As a result, civil servants, administrators, and politicians took it upon themselves to "spread the word" in turn. They welcomed opportunities for planting and holding trials for new eucalyptus species.[27]

Eucalypts in Victoria

The Australian state of Victoria, and notably Melbourne, its capital, proved to be an important location for early research, experimentation, and seed shipments. Working as government botanist in the second half of the nineteenth century, Ferdinand von Mueller addressed the benefits of using eucalypts at home and growing them overseas, especially in warm temperate and subtropical nations, where he established many professional contacts.

Melbourne's Royal Botanic Gardens served as the site from which Ferdinand von Mueller pioneered and promoted eucalypt research and exchange for forty years.

Cottage on the grounds of Melbourne's 35.4 ha (87.5 acre) Royal Botanic Gardens, which currently display more than forty-nine thousand plants.

Eucalypts Overseas

A dozen or so species, noted for speed of growth, adaptability, coppicing, and uses as poles, charcoal, fuel, and pulp, occupy the largest amount of land area planted under Australian eucalypts in foreign countries.

E. camaldulensis (river red gum) is probably the most widely planted of eucalypts, growing in Spain, North Africa, Southeast Asia, and many parts of South America. It grows on infertile and dry soils, and tolerates drought and high temperatures. This specimen is in Melbourne's Royal Botanic Gardens.

E. saligna (Sydney blue gum) is scattered along coastal areas of east and southeast Australia. Excellent for pulp and coppicing, together with *E. grandis* (flooded gum), it is a favorite plantation tree in Brazil.

E. viminalis (manna or ribbon gum) tolerates frost, moderate fire, and coppices well as a plantation species. South Africa (in the past), Spain, Chile, and Argentina (recently) have proven major outlets in more temperate regions.

E. grandis, another east coast species in Australia, grows rapidly as an industrial wood, especially in Brazil and South Africa.

E. tereticornis (forest red gum) has flourished in India (where it has come to be considered an indigenous tree), Africa, and South America. It is highly regarded in plantations, surviving in humid tropical sites with summer rains, and furnishes a range of wood products. This California stand was probably photographed around 1900. (McClatchie, *Eucalypts Cultivated in the United States*)

Gum tree bark (from California) shows the special tendency to flake and shed. Spent leaves and bark plus low levels of light may inhibit growth of an understory in overseas areas where trees are planted close together.

California Promotes Eucalypts

Up to the present time knowledge based on
actual utilization of California-grown
eucalypts is meager. Not only has no large
amount of timber of merchantable size
ever been available, but manufacturers and
consumers are naturally reluctant to use
new and little-known wood in place of those
which have proven entirely satifactory for
certain purposes.   —H. S. BETTS AND
                    C. STOWELL SMITH, 1910

The six decades after 1830 were the heyday for initiating crude trials with
eucalypts in new regions. While von Mueller and colleagues promoted the
plant genus in Australia, collectors and officials in approximately fifteen additional nations were conducting trials with numerous species. Young trees were
shooting up in sheltered localities in the United Kingdom and Ireland, in
southern Europe, North Africa, and India. Across the Atlantic Ocean, eucalypts were growing successfully in Chile, Brazil, Uruguay, Argentina, and Peru
in 1860 (see appendix 2, table 2). About the same time, promoters and nurserymen in the United States were beginning to take seriously these novel
plants, intended for shade and ornament.

Interest in California, the "home" for eucalypts in the United States,
coincided with the Gold Rush. Reportedly, San Francisco immigrant Hans
Hermann Behr, who had worked with Australian eucalyptus and acacias in
Australia, extolled their value in the New World. The German-born Behr, who
had studied under Alexander von Humboldt, had visited Australia twice and
befriended von Mueller. He began to cultivate eucalypts in about 1850. Three
years later, fourteen species of young eucalypts were speeding skyward in a

San Francisco nursery owned by Willam C. Walker, to whom Behr may have given seeds. In 1857 Walker ran an advertisement for the new plants in the *California Farmer* and exhibited gum varieties in San Francisco's Mechanics Fair. About the same time, young gum trees were growing in an Oakland nursery across the Bay, where they sold for 5 dollars apiece.[1]

Early in the next decade Stephen Nolan, who operated the Bellevue Nursery in Oakland, obtained seed stock directly from Australia. Nolan had gotten his start by working on the estate of George Potter near Oakland and was probably familiar with Walker's Golden Gate Nursery across the Bay. He marveled at how fast these eucalyptus grew and liked their toughness, so he contracted a sea captain to bring back seeds from Australia. Nolan raised and sold large numbers of eucalypts throughout the state. By 1871 his catalogue listed thirty-four species, selling for 50 cents or less apiece.[2]

Initial experiments with the Australian trees met two complementary objectives. The first was a growing demand for plants as novelties and money-spinners. Rich Californians purchased land, constructed homes and estates, and filled them with familiar and unusual plants, including eucalypts, which looked attractive and added shade quickly. Commercial nurseries proliferated in order to serve this booming market for ornamental and novel species. James T. Stratton, surveyor-general of California, pioneered blue gum cultivation in Hayward, Alameda County. He obtained seed from Bishop William Taylor's wife Annie, who had begun growing them after her husband sent back seed from Australia while on a religious mission. Reportedly, Stratton set out 45 hilly acres (18 ha) with eucalypts and later marketed tens of thousands of seedlings.[3]

Growing eucalypts satisfied a second interest. The influx of population during the Gold Rush created demand for local wood for construction and fuel. Consequently, citizens grew concerned about rapid deforestation in and around the Bay Area. In 1858 Colonel James Warren, an acquaintance of Walker, published an article in the *California Farmer* (an organ that he owned and operated) about the value of planting trees "to supply the places of those which have been consumed . . . for fires, and artistic and mechanical purposes." Some authorities credit Warren with distributing blue gum seeds beginning in the mid-1850s in order to restore timber trees to California.[4]

The issue of replanting came to a head in the following decade when, in 1865, Rev. Frederick Starr predicted that within thirty years the state and nation would have to face a timber famine. Three years later the California Tree Culture Act of 1868 encouraged county supervisors to plant trees along roads. In 1873 the federal Timber Culture Act also responded to concerns about deforestation by requiring homesteaders to plant 40 acres (16 ha) (later reduced to 10 acres [4 ha]) in trees, for every 160 acre (65 ha) claim. It was no surprise, therefore, that people singled out eucalypts as virtual godsends. Author Robert E. C. Stearns argued that the blue gum, which was by then becoming fairly well known and distributed, was the obvious tree for farmers to plant and for investors to market. In the late 1860s and early 1870s, prominent officials and public figures began to praise eucalypts for their utility and beauty, and encouraged investors to expand cultivation.[5]

At that time, horticulturists elsewhere were speculating about cultivating these and other foreign plants. In Texas, for example, entrepreneurs planned to grow coffee and tea along the Gulf Coast. They argued that Asian-derived orange trees would thrive in warmer counties next to the Gulf of Mexico. Plums and peaches prospered on the limestone uplands of the central Texas hill country. In March 1875, at a meeting of the state's earliest and most famous horticultural society in Cat Spring (east-central Texas), a member proposed that *E. globulus* would flourish in south-central counties. These unusual Australian trees, he declared, would lend the atmosphere an aromatic and, therefore, healthful quality and prevent fevers.[6]

Unlike California, however, Texas failed to become a reliable home for eucalyptus. Frosts killed and injured stands, even in the balmiest sections of the lower Rio Grande Valley. Nonetheless, Texas was one of the three states (Arizona and Florida were the others) for which eucalypt experts still predicted a bright future, in addition to California. Promoters tried them unsuccessfully as far north as Fort Worth, but achieved the best results along the corridor of the St. Louis, Brownsville & Mexico Railroad that wound into Brownsville. For thirty or forty years the warmest part of south Texas boasted groves of eucalypts. Tens of thousands of trees representing more than thirty species sprouted as ornamentals and windbreaks. Farmers set them around orange groves in the Rio Grande Valley. All told, some twenty or so cities and small towns from Eagle Pass eastward to Beeville and Tivoli, and from San Antonio south to McAllen and Brownsville boasted eucalypts.

Eucalyptus rostrata (river red gum, now known as *E. camaldulensis*) proved most popular because it withstood heat and cold, and took to saline soils. However, much to the dismay of nurserymen, hard frosts, drought, and alkaline soils gradually picked off the trees, eventually souring public support for this gum species and others.

In the early 1900s, when the boom in both California and Texas was at its height, glowing reports about the trees being impervious to insects, making fine wood, towering majestically, and "promoting healthfulness" added to the fame of eucalypts. But as years ticked by and prospects for profits dimmed, Texans turned to other things. A few plantsmen who had worked and traveled in Australia continued to grow eucalyptus. In the late 1950s, for example, eucalyptus trees shaded the streets in Carrizo Springs, after Ernest Mortensen, superintendent of the nearby state Agricultural Experiment Station, planted them. Mortensen had worked for the Australian government, and he planted in that Texas community seeds of approximately three dozen species he had imported.[7]

Initial Commitment

A small number of influential and wealthy citizens promoted the introduction of *Eucalyptus* and pointed toward the industrial, agricultural, ornamental, and health benefits to be reaped from the cultivation of different species. Two men in particular established themselves as experts and popularizers in central and southern California. One was Ellwood Cooper, educator and president of Santa Barbara College, who had first glimpsed the Australian trees on an initial visit to the Bay Area in 1868. Several small woodlots near San Francisco, planted a dozen years or so earlier, caught his eye. Cooper decided to experiment with the unusual trees when he settled in Santa Barbara two years later. He wrote the first book published in the United States about eucalypts (*Forest Culture and Eucalyptus Trees*, 1876) and worked assiduously to spread them. The second spokesman was Abbot Kinney, a real estate magnate, land speculator, and wealthy developer in Los Angeles, who championed the cause of eucalyptus among state and federal forestry agencies as well as to the public. Kinney published the second important American book about eucalyptus (*Eucalyptus*, 1895). It is an authoritative text, which details a range of economic, health, and environmental benefits to be gained from cultivating various species. His book appeared twenty years after Cooper's.

Cooper was one of a growing group of enthusiasts interested in experimenting with new, interesting, and potentially useful plants. He dedicated about 200 acres (81 ha) of his ranch to eucalypts. Struck by their graceful shape and silvery, pendulous leaves, in the early 1870s Cooper used these Australian plant immigrants to line the grass-covered hills of his Dos Pueblos Ranch outside Santa Barbara. Contacts with an old acquaintance, Thomas Adamson, the U.S. consul-general in Melbourne, Australia, heightened Cooper's regard for the trees. Adamson sent him seeds, and within a few years Cooper was growing thousands of eucalypts, many of which sprang up with extraordinary rapidity. One tree, claimed Cooper, reached 12.9 m (42.3 ft) high with a trunk 24.1 cm (9.5 in) in diameter in barely six years.

Consul Adamson also put his friend in touch with Ferdinand von Mueller, whose guidance Cooper sought in regard to the procedures and techniques required for successful cultivation. From Melbourne, von Mueller supplied his new correspondent with the appropriate information, requesting that Cooper recompense him by making his botanical research and opinions about eucalyptus better known in the United States. Cooper agreed and arranged to publish a number of von Mueller's lectures about the value of *Eucalyptus*. These formed the body of Cooper's book.

In *Forest Culture and Eucalyptus Trees* Cooper suggested that arboriculturists should compare California with Australia because "climatic and other conditions" for growing trees are very similar. He included von Mueller's testimonial about the value of forest trees for the health of humankind, and his remarks about the need to set aside botanical gardens in which scientists would be able to experiment with, conduct research on, and promote various plants. The bulk of the 621-page text consists of materials sent by von Mueller. These gave details about the natural history and uses of eucalypts and referred to trials in Australia.[8]

Threaded throughout the book is the firm belief that humans can and must improve upon nature. Von Mueller explained how he had cultivated nonnative as well as native plants, importing to his botanical sanctuary a range of herbs, shrubs, and trees from North America and Asia. He included, for example, tea plants, for which he anticipated a great future. Exporting Australia's best and most useful plants (and animals too) was part of this contract for improving upon nature. It was also an economic service one would expect from a servant of the British crown.

Although he understood and shared von Mueller's esteem for the material value of eucalypts, Cooper struck a more aesthetic tone. In 1875 he delivered a "milestone" lecture on eucalypts, declaring that "at my home I have growing 50,000 trees." The open, rolling hills around his Santa Barbara ranch were taking on a variegated aspect, Cooper explained, so that as woody trees spangled the native grasses, the landscape itself grew more interesting and pleasing.[9]

Cooper took special pleasure in declaring how beautiful many parts of southern California were becoming, as the fragrant woody plants supplanted plainer native grasses. Urbanites as well as ranchers were taking to the species. In 1874 the U.S. commissioner of agriculture reported that people had planted at least 1 million blue gum trees, "in streets of principal cities and in country localities where the winter is sufficiently mild." San Francisco mayor Adolph Sutro encouraged local children to set out eucalypts as part of Arbor Day observations, targeting the Presidio, Mount Davidson, and Yerba Buena Island.[10]

During the 1870s and 1880s promoters argued that eucalypts would provide a vital asset for California and the nation at large. As foresters endeavored to overcome a nationwide shortage of usable timber, notably hardwood, they embraced eucalypts. Cooper believed that gums—a thousand to the acre—were an excellent remedy for the ongoing depletion of native hardwoods. He managed and maintained his woodlots carefully, believing that "forest cover" had many advantages. In addition to proving themselves a valuable source of timber and fuel, eucalypts acted as windbreaks and shaded rural roads and city streets. Eucalypts figured in plans for estates and gardens, furnishing "an object lesson of what the tree will do for an appreciative planter," declared Cooper. Thirty or so years after he had started, Cooper maintained stands of eucalypts; visiting foresters considered his estate "the best place in America to see a large variety of Eucalypts grown as forest trees."[11]

Millionaire Abbot Kinney settled in California in 1880 and founded the city of Ocean Park. He became chairman of the California Board of Forestry and built on Cooper's trials and wood practices in Santa Barbara. He also drew upon the recondite knowledge of people like von Mueller and tree scientist Alfred J. McClatchie, who authored a bulletin about eucalypts in America. McClatchie and Kinney became friends and worked closely together to popularize the Australian plants. McClatchie admired Kinney as an "enthusiastic student and planter of trees of this genus." Kinney used his official position in California forestry to broaden the range of sites for trials, increase species numbers, and make eucalyptus trees popular. He pushed for expansion, especially of *E. globulus,* throughout the American Southwest. This practical innovator penned "the most extensive American work on the genus . . . from the aesthetic, the botanical, and the utilitarian standpoints," opined McClatchie in 1902.[12]

Kinney's authoritative book *Eucalyptus* appeared in 1895. In it he explored the range of uses for gum trees and explained methods of growing them. Even after almost fifty years of trials in California, Kinney declared, there was still relatively little detailed information from growers in the United States. Americans needed to know what species to plant, and where and when to plant them. His book sought to rectify this omission. It drew upon information and suggestions provided by accomplished foresters like Cooper at home and von Mueller abroad.

Kinney, who was well connected, widely read, and influential in scientific circles, particularly in southern California, agreed with Cooper wholeheartedly about the benefits of eucalyptus, including the capacity of some species to self-seed and hybridize, characteristics more like those of native plants, and which could be turned to material advantage, he argued. Kinney distributed seeds and used forester colleagues to circulate seedlings to the general public. For example, the state Forestry Commission, with which he was connected, sold seeds cheaply. The University of California at Berkeley also gave away eucalypts to growers. Kinney served as chairman of the Board of Forestry between 1886 and 1888. Under his direction, the board received land on which to build experiment stations. One of them, in Santa Monica, Kinney's home, came to possess an impressive eucalypt collection and reportedly distributed 76,000 seedlings to 421 growers in 1890 alone. The University of California's College of Agriculture in Berkeley took control of operations, and under the direction of Dean E. W. Hilgard, it pressed ahead with eucalypt research and made seeds and seedlings available to the public. By the turn of the century, more than one hundred fifty species of eucalypts had been tested and grown in various locations. Kinney noted that in a day's walk around Berkeley an observer would come upon a dozen or more eucalypts, including several tall, handsome groves of blue gum on the university campus.[13]

Testimonials flooded in. Well-known people admired the trees and wanted more of them. Cooper sent seeds to his friend, Judge Charles Fernald, who passed them to others. Author Jack London, for instance, reportedly set out a hundred thousand eucalypts on his estate on Sonoma Mountain. Railroad companies, including the Central Pacific, planted rights-of-way with eucalypts, anticipating—incorrectly, as it turned out—a future market in ties and fuelwood. The fine-grained wood of eucalyptus trees supplied decorative finish. Sales of eucalyptus oil and honey grew.

Frank C. Havens, an Oakland developer, initiated a profitable venture to clothe a 14 mi (22.5 km) stretch of the East Bay skyline from Berkeley through Oakland in one ribbon of eucalyptus. Between 1910 and 1913, Havens ordered 8 million trees planted in the rolling hills, intending to sell them off eventually as lumber. He constructed a sawmill. Results, however, disappointed him. Research by this speculator and by U.S. Forest Service personnel concluded that blue gum less than 2.5 ft (76.2 cm) in diameter was unsuitable for milling. Wood from the young trees bent and split too readily as it dried. In Australia, foresters felled gums for timber when they were decades or even centuries old. In California, weedy, immature gums, such as those growing in the Berkeley hills, proved useless. Timber experts rejected them, and Havens's mill folded.[14]

Despite this setback, promoters of eucalyptus insisted that gum wood could be made into first-class timber. They looked for the new Panama Canal to open markets by directing shipments to destinations along the Gulf and Atlantic seaboards. Recognizing a good thing, nurserymen took their cue, expanded purchases of seeds and seedlings, and pushed retail sales. Real estate developers claimed that buying land for eucalyptus plantations was a money-maker (after all, Kinney was an astute businessman). This get-rich mindset led to a boom in the early 1900s, which archivist Robert Santos has called a "full blown eucalyptus industry in California." He reports at least one hundred companies selling land, managing nurseries, and planting trees.[15]

Institutional Support

In the last two decades of the nineteenth century, alien eucalypts began quite literally to change the face of central and southern California. Kinney's friend McClatchie—the agriculturist and horticulturist Alfred James McClatchie (1861–1906)—worked for the U.S. Department of Agriculture Bureau of Forestry and was a faculty member at Throop Polytechnic Institute, Pasadena. McClatchie wrote that "eucalypts serve as a forest cover to mountains, hills, plains, and swamps, as wind-breaks, and as shade trees. While growing they are also the source of many gums, resins, and of honey. When cut, they furnish valuable timber, excellent fuel, and a very useful oil." Taller and taller groves were greening up sections of Arizona, New Mexico, Texas, and Florida, he noted. Except for extremes of heat (that is, midsummer temperatures in excess of 39°C) and cold (with winter lows down to –10°C), the new geography of gum trees had no limits. They blossomed in areas with warm summers (27–37°C), cool winters (generally above freezing), and seasonal rains. On a national map, this translated into a belt of country 100–300 mi (160–480 km) wide that ran along the nation's southwest and southern sections. Eucalypt "habitat" started in north-central California, including the Central Valley, extended to the Mexican border, turned east into Arizona, New Mexico, and Texas, and hugged the Gulf of Mexico through southern Florida.[16]

The "optimum" zone, promoters declared, was in southern California, southern Arizona, lower New Mexico, and down the Rio Grande to the Gulf of Mexico. The eucalyptus zone ballooned into central Texas, then dropped back to the coast. Curiously, Kinney was not as confident as McClatchie about this predicted range. Kinney noted that even the "fastest growing tree in the world," blue gum, had failed in parts of Arizona and Texas. Searing heat and excessive frosts threatened its geographic empire. He believed, however, that other species of eucalypts would compensate and adjust for blue gum losses.[17]

McClatchie, in contrast, had every expectation that eucalypts would blanket the American Southwest. He exclaimed with unscientific exuberance that eucalypts would "probably serve more useful purposes than the trees of any other genus grown on the globe" (with the possible exception of palms). Optimism marks McClatchie's 106-page work, *Eucalypts Cultivated in the United States* (1902), an authoritative treatment of the plant genus in the United States.[18]

McClatchie's work, published by the U.S. Forest Service, reflects the agency's interest in useful woody plants. Beginning in 1887, with a report on railroads and forest supplies, the Forest Service sought to balance timber output with demands for wood products. Scientists surveyed timber supplies in the Rocky Mountains, Adirondacks, and Cascades, and in Wisconsin, Arkansas, and Puerto Rico. They discussed planting trees on the high plains. Early publications expressed the agency's commitment to make sure there was enough timber to meet the nation's needs.

McClatchie's lavishly illustrated work was a departure. It was the first detailed discussion about planting an alien tree genus. Earlier essays had explored the systematics of white pine, western hemlock, and red cedar, and had included timber pines for the South. McClatchie's work on eucalypts showed how the Bureau of Forestry (as it was then called) was casting a broader net, involving itself in woody plant introductions.

McClatchie's Forest Service bulletin is detailed and comprehensive, and stands out among a dozen or so bulletins, circulars, pamphlets, and short books that federal and state scientists devoted to eucalypts during the first years of the twentieth century. These forester-authors offered veritable handbooks for planters. They told people how to obtain and plant seed, and informed them about diseases. They named researchers and listed additional readings. They described what investors could expect from the cultivation of various eucalypts (McClatchie detailed forty-one advantages). McClatchie, who worked out of Phoenix, reported how different eucalypts adjusted to different conditions, especially in California, where soils, precipitation, and temperatures ranged from cool and moist in the north to hot and dry in the Imperial Valley and Mojave Desert in the south. Such climatic shifts, plus an amplitude in relief—from sea level to the Sierra Nevada foothills—affected tree survival. Drawing from experiment station data, foresters eagerly passed along recommendations to growers throughout the state.[19]

McClatchie had an extraordinarily high regard for the Australian trees. His support for them trumpeted the usual virtues, for when his bulletin appeared the "boom" was still on. McClatchie, other scientists, entrepreneurs, and interested nurserymen anticipated making sizable profits from growing and selling eucalyptus.

The alien trees would produce valuable forest cover, which was especially important for stabilizing slopes and shrouding dry, exposed soils in what McClatchie termed the "the treeless land of semitropic America." He drew from correspondence with Ellwood Cooper (dated June 1900) to report how eucalypt seedlings had stabilized soils on a hillside near Cooper's Santa Barbara ranch after a rainfall of 14 in (35.5 cm) had battered the slope for several days.[20]

McClatchie claimed that eucalypts made powerful windbreaks. He described how fruit growers in Santa Paula, for example, used rows of *E. globulus* to shelter their orchards. Without such protection "large areas in Ventura County, Cal., would not be tillable," he advised, due to blowing soil. The windbreaks reduced wind damage and helped protect against frost. In this regard, McClatchie judged eucalypts superior to cottonwoods; they also gave year-round shade over pastures and roads.[21]

McClatchie did not say that eucalypts affected other plants growing around them or that they drew soil moisture from orchard crops. He glossed over such disadvantages in his zeal to emphasize how quickly eucalypts speed upward, thus minimizing wind and frost damage. Shade was the chief reason why farm advisers in five southern California counties continued to promote eucalypts long after McClatchie, and why the trees remain a conspicuous woody feature in many areas.[22]

McClatchie turned to timber, repeating von Mueller's point that there were at least twenty-five special uses for eucalyptus wood in Australia. Craftsmen in that country turned the wood into posts, planks, and farm implements, and were now doing the same in the United States. Cooper had sold more than $10,000 worth of eucalyptus for pilings over the previous decade, noted McClatchie. He added that stout posts and piles of these exotic woods proved more durable for piers along the Pacific Coast than either redwood or Douglas fir.

Colleagues revised McClatchie's claim twenty years later. California Agricultural Experiment Station official Woodbridge Metcalf insisted that native conifers, such as Douglas fir, were at least as resistant to decay as untreated blue gum. Blue gum suffered worm damage in saltwater. Large pilings made from any eucalypt other than blue gum were hard to find, Metcalf noted, and either proved too misshapen or else were not available in quantities sufficient to merit marketing. Experiments revealed that posts made from small eucalyptus trees rotted quickly as the sapwood gave way. Metcalf concluded that "the wood of young trees is not durable when in contact with the soil."[23]

Although McClatchie admitted that eucalypts had been cultivated neither "long enough nor extensively enough to have become a major source of lumber" in the United States, he had no doubt that the trees would "become sources of much timber." It was just a matter of time. His position raised two problems. First, McClatchie and his associates assumed that the many and varied uses for these hardwoods in Australia could soon be replicated. Although he admitted that variations in strength, durability and flexibility occurred among species, and that differences in edaphic and weather regimes affected wood properties, McClatchie never questioned the premise of being able to transfer felling, curing, and carpentry practices from Australia to North America. In addition, he seemed to envisage that the range of manufactured products would be the same.

In regard to the question of wood products, McClatchie avoided the sensitive issue of immature wood. Fast-growing eucalypts in California furnished inferior timber compared with mature trees in Australia. Six years earlier Kinney had faced this same problem of the warping and splitting of recently cut wood. Kinney admitted that the Southern Pacific Railroad Company had experimented with ties made from blue gum, only to find that they "checked [cracked] to such an extent that room could hardly be found to bolt down the rails." In pages devoted to gum trees "as a source of timber," McClatchie stressed the range of benefits but none of these difficulties.[24]

It was a major oversight. "Very little lumber has been made from California Eucalyptus trees except in an experimental way," admitted Metcalf, due to "excessive checking and warping" during seasoning. Although eucalyptus wood was hard, logs cut from ten- or fifteen-year-old trees cracked, split, buckled, or shrank as they dried. Wood in Australia was usually older and easier to process, but Metcalf concluded that the younger woods in California were unsuitable for anything but smaller items, such as toy parts, handles, and insulator pins, for which a market existed.[25]

"The trouble with California Eucalyptus is almost invariably the wrong species have been planted," declared H. D. Thiemann. Forester Thiemann carried out seasoning experiments with blue gum at Berkeley from 1912 to 1914 and later traveled to Australia for further tests on mature trees. He recommended that eight species, then mostly rare or absent in California, be given trials. Most, such as ironbark, jarrah, and gray box, furnished hard and durable wood but grew more slowly than blue gum. Like blue gum they could not take hard frosts. However, Thiemann knew that a few specimens existed on federal or state lands, and he hoped that further study and trials would make them popular. In his view, the blue gum was "excellent fuelwood" but not "lumber at an age of less than 75 to 100 years." Similarly, none of the four most populous species in California—namely, blue gum, red gum, gray gum, and sugar gum—had much timber value. The glamorous aura that McClatchie had cast on the genus had dulled considerably.[26]

Fuelwood was another major reason for planting eucalyptus. Drawing liberally upon Cooper's reports and correspondence, McClatchie noted that blue gum made excellent fuel. Other equally fast growers, such as manna gum (*E. viminalis*) and red ironbark (*E. sideroxylon*), which grew more slowly but reportedly burned better, also proved superior stove woods. All three eucalypts could be cut on a five- to seven-year basis and would then coppice or sprout back.

McClatchie made no mention of a shift from wood to oil for fuel. He did not acknowledge that wood was being phased out for heating, cooking, and locomotion, and he did not point out any potential drawback in putting some of California's best cropland into 4–16 ha (10–40 acre) groves of eucalypt fuelwood. McClatchie regarded such plantings as profitable investments. They also relieved the monotony of the open scenery.

Cooper, noted McClatchie, chopped 1,000 cords annually from his eucalypts without making an overall dent in the forest cover. In addition, the Santa Barbara innovator grew many trees on soils too barren for agriculture. Since there was no dearth of desert-like places, McClatchie opined that "the Eucalypt is evidently destined to be the future fuel tree of the Southwest," and that therefore an extensive drive to reforest with eucalypts "would be a wise provision for the future."[27]

By the early 1920s others had moved away from McClatchie's position. True, the wood burns brightly and gives off a fragrance. Superior fuelwood trees are about ten years old and burn as well as native species. Marketing, however, was the problem, especially when hauling logs over distances. Then, native wood competed with eucalypts, except as charcoal. Metcalf's research into eucalypt plantations for fuel and charcoal reported a weakened economic viability. Breakthroughs that subordinated wood to petrochemical energy had increasingly marginalized fuelwood.[28]

McClatchie relied on Kinney for information about honey and oil. Around Kinney's home in Santa Monica, where eucalyptus trees dated back to 1876, lofty blue gums flowered from January to May. In spring, these "bee pastures" supplied excellent honey, noted Kinney. His friend reported that beekeepers had grown increasingly excited about eucalypts as sources of nectar, and listed thirteen species as excellent suppliers of honey.[29]

Oils formed part of McClatchie's comprehensive survey. Pioneer work in Australia, bolstered by research in North Africa, proved that plantations provided enough oils for medicines and perfumes. In California, blue gum supplied eucalyptol for soaps and cough medicines at a rate of 9 tons (8.2 metric tons) of oil from 700 tons (635.5 metric tons) of leaves and twigs. A company called California Eucalyptus Works in Garden Grove specialized in extracting oil from *E. globulus*. Reportedly, a Los Angeles physician was attempting "to establish a reputation for putting up a pure high-grade product."

The antiseptic qualities of eucalypt oils were widely noted, said McClatchie, who reported that the "oil is also a well-known remedy for malarial and other fevers, and is used in treating diseases of the skin, and of the stomach, kidneys, and bladder, and is especially valuable for affections of the throat, bronchi, and lungs." Local remedies included steaming leaves of blue gum for colds and catarrh, and brewing tea from them.[30]

In the early 1920s about 150,000 pounds (68,000 kg) of eucalyptus oil, valued at $75,000, entered the United States over a span of several years, mostly from Australia. Unfortunately, its type and quality were superior to oil manufactured in California. Metcalf noted, for example, that blue gum oil did not meet U.S. pharmaceutical standards without expensive redistillation. He argued that "further investigations are necessary" to establish the methods and equipment necessary to generate high-grade oil. At that time, limited quantities went into "unofficial" medical preparations, soaps, boiler compounds, and flotation oils. Extracts found outlets as antispasmodics, disinfectants, sedatives, and decongestants.[31]

Disease

For many, the decisive, though controversial, allure of eucalypts was their capacity to suppress malaria while also improving local climate. McClatchie reviewed the general belief predominant in Mediterranean Europe and North Africa: that the Australian trees had a "distinctly sanitary" effect on climate and on the areas in which they prospered. He referred to the all-too-commonly mentioned case of the Tre Fontane monastery near Rome, explaining how Charles Belmont Davis, American consul in Florence, had cited it as evidence of the capacity of the genus to contribute a "balsamic atmosphere" and thereby improve human health. McClatchie omitted disparaging remarks made by Davis, arguing instead that affection for the tree was prompted by its capacity to dry out marshy places and render the climate more salubrious. In this respect, McClatchie repeated the conventional wisdom that Californians, like the monks in Italy, had eagerly turned to the "fever trees" in order to suppress agues.[32]

In the thirty or so years between the first claims about the importance of eucalyptus for human health and McClatchie's 1902 Forest Service bulletin, debate intensified about medicinal uses of the genus. McClatchie acknowledged the strong disagreements about the claimed links of eucalyptus to health and climate, and took a middle ground. He stated soberly that "it is entirely reasonable to believe that to a certain extent they beneficially affect the atmosphere in the region of their growth." This was because eucalypts absorbed ground and surface moisture, and the "exhalation" of volatile oils from foliage and "purification of germ-infested matter" from fallen leaves left wooded landscapes disinfected. McClatchie was vague about documenting such sanitary proclivities. He grew defensive, declaring that there was no harm in thinking that eucalypts ameliorated climate even if they didn't. "That they improve climate has served a useful purpose regardless of the facts of the matter," he concluded timidly.[33]

McClatchie's position reflected that of his friend Kinney, who reprinted a series of older medical articles about the health benefits of eucalypts and their essential oils. Citing *Pharmacology of the Materia Medica,* for example, Kinney noted that preparations made from oils or extracts, such as elixirs, lozenges, inhalants, and pills, filled a range of needs. Some were stimulants, aphrodisiacs, antispasmodics, or antiseptics, and were used "in the treatment of intermittents, especially in those chronic varieties in which quinine has failed."[34]

Taken internally—a dangerous proposition—a eucalyptus potion may not act as quickly as quinine, Kinney admitted, but medical experts judged it better for severe cases, especially where patients were chronically exposed to malaria. Also, eucalyptus extract was cheap. A tincture could be made by covering 3 oz (85 g) of crushed fresh blue gum leaves with 6 oz (177.5 cc) of alcohol, then allowing them to sit for two weeks in a closed vessel. The extract was then decanted and filtered. The resulting fever suppressant reportedly ranked second only to quinine.[35]

Dr. H. A. Foster admitted in the *Therapeutic Gazette* (1880) that although there was no "one uniform and universal application" for malarial fever, he had confidence in eucalypt tincture. In fact, for about a year he had regarded it as an "unfailing remedy," until several preparations had failed to arrest so-called malarial poisoning. Foster supposed that, like other remedies, the eucalypt tincture had effects that varied "in respect to particular seasons and localities for its better action." He noted, however, that it had acted when quinine had failed.[36]

Similar reports, reprinted by Kinney, expressed essentially the same message. When ingested in various forms, eucalyptus soothed and healed patients. It proved useful for catarrhal afflictions, venereal disease, bronchitis, bladder infections, cystitis, diphtheria, typhoid fever, and dysentery. Its juices imparted warmth and facilitated digestion. Eucalypt extract proved a useful antiseptic in surgery. One report noted that it relieved the fetid smell from the gangrenous leg of a patient. Gangrene had spread too high up for amputation, but a salve of eucalypt gave relief, and with continued dressings the man remained free of odor "until death occurred, two or three weeks subsequently."[37]

Kinney posited that malaria was caused by a germ introduced by air and water. Unboiled water from shallow wells was one way of contracting "ordinary American" malaria, he insisted. Eucalypts arrested this bacillus as their roots drained stagnant swamps, and their leaves disinfected standing water. Extensive plantings in Italy, Spain, the south of France, North Africa, India, South Africa, Argentina, and other places had markedly improved the health of the local populace, "especially as to malaria," he stated.

Drier, hygienic soils were one set of benefits, leaves that wafted aromatic odors were another. They filled the air with antiseptic vapors, supposedly suppressing wind-driven bacilli. In concentrated form, Kinney declared, these essential oils were "fatal to all insect and bacterial life." Just as a eucalypt concoction had apparently given relief to the man with a gangrenous leg, so could a grove of the trees disinfect the land surface, its soils and waters. Kinney noted that mosquitoes avoided such disinfected spots; the foliage repelled them. In other places, however, specifically in certain woodlots along the Santa Barbara coast, he admitted that mosquitoes were "liveliest and most savage."[38]

The De-adoption of Eucalypts

Six years after McClatchie's publication appeared, Norman Ingham, a forester associated with the University of California Agricultural Experiment Station in Berkeley, published an update. Ingham, foreman of the university's Forestry Station in Santa Monica, where seventy species of eucalypts were being tried, suggested that "planting has passed the experimental stage and may be considered without question as a commercial proposition, particularly in regards to wood." While admiring the trees and repeating many of McClatchie's upbeat depictions, Ingham noted ominous signs in California. Entrepreneurs and developers were having doubts about the long-term future of eucalyptus in California.[39]

In respect to numbers, things seemed to be progressing well. Ingham knew of about a hundred species in California, mostly intended for lumber and fuel. Still, he admonished readers that it would be "unprofitable and unwise to enter upon Eucalyptus planting with the sole idea of raising wood for fuel." Why? Because "the production and use of natural oil is rapidly increasing." Cities were turning to cheap gas and electricity, so that fuelwood was "rapidly becoming a luxury." Land was so valuable for other crops that it did not make sense to plant for fuelwood. Under irrigation, marginal areas that had previously gone under the alien trees could be planted with horticultural produce instead.[40]

Nevertheless, Ingham believed that profits could be made from the trees as timber. Noting that in 1907 the Hardwood Planing Mill in San Jose had "sawed up many thousands of feet," mostly from blue gums thirty years old or older, he judged that the finished products were as valuable as oak. Ingham was probably referring to an establishment run by one J. T. Gillespie, who specialized in processing "San Jose Blue Gums." These trees had likely originated from seed gathered in Victoria, not Tasmania, and which could be made into quality timber, depending on the time of year the trees were cut and the length of seasoning given them. Gillespie matured his timber in open sheds for about two years before fashioning the best wood into beams, wagon beds, frames for hayrakes, and even furniture. But few others seemed prepared to invest in eucalypts for timber. There was a downside, Ingham admitted. Eucalyptus wood was costly to cut and mill, and when it was being planed, it tended to chip due to an irregular grain. It was also expensive lumber.[41]

When Ingham penned his report, stretches of California from Ukiah southward along the coast and inland in the Central Valley from Redding to Bakersfield carried eye-catching stands of eucalypts. Blue gum grew on the coastline and along river valleys in which average rainfall exceeded 15 in (38.1 cm) per year. It was an adaptable species, usually able to withstand cold winters as far north as Red Bluff and dry, hot summers in the San Joaquin Valley. "Nearly every town south of San Francisco to San Diego," declared Ingham, boasted a grove or two of gum trees, even approaches to the furnacelike Imperial Valley. In the valley itself, planters counted on two species: river red gum *(E. camaldulensis)* and West Australia's flooded gum *(E. rudis)*. These adapted best to desert aridity and heat.[42]

Gradually, more and more people admitted that profits were not as bountiful as men like Cooper and Kinney (and their distant mentor Ferdinand von Mueller) had predicted. Railroad companies grumbled that even treated eucalyptus wood made only second-rate track ties because so much of it rotted or cracked. Among the few hundred ties laid by the Southern Pacific in central Nevada, some failed to hold any spikes and others split badly, lasting only a few years before they fell apart. In the late 1880s and early 1890s, three U.S.

Forest Service bulletins had begun to challenge this use for eucalypts by discussing how materials other than wood could be used as cross-ties.[43]

In December 1910 two foresters, H. S. Betts and C. Stowell Smith, disparaged previous claims for gum woods, insisting that Forest Service data had "at times been misused." In their Forest Service circular "Utilization of California Eucalypts," the assistant district foresters noted that "extravagant estimates of the probable returns from planted eucalyptus have been widely circulated." They continued, "There is reason to fear that many persons have formed an altogether false idea of the merits of eucalyptus growing as a field of investment." Such persons might blame the Forest Service for misleading them. Betts and Smith's thirty-page publication was intended to be a warning. "There are as yet too many elements of uncertainty" in regard to monetary returns, the authors insisted. They intended to set the record straight, especially in respect to the value of wood "for high-grade purposes." And how valuable was eucalyptus wood? Not as valuable as growers had hoped, due to the tendency of the immature timber to warp, shrink, and check.[44]

Collaborative experiments with the University of California showed that eucalypts were not suitable replacements for the waning supply of hardwood (mainly native oak) on the West Coast, as people had anticipated, noted Betts and Smith. The absence of distinctive bands denoting spring and summer growth made it difficult to age eucalypts. But that was a minor problem compared with discovering that in some species the wood fibers were usually interlaced, which made the logs difficult to split. An irregular grain of the wood led to chipping, and the large amount of water in newly felled trees made logs very difficult to season and so heavy that some sank rather than floated in water. It was this seasoning, or drying, that caused most trouble. In drying, blue gum shrank as much as 22 percent (compared with 12 percent for white oak), and experts had been unable to devise methods for reducing the splitting, twisting, and checking, despite perusing Australian publications. The best that Betts and Smith could say was that "California-grown eucalypts have shown themselves practically equal to certain of our native woods . . . but no satisfactory method of seasoning eucalyptus lumber grown in California has yet been worked out on a commercial basis." This puzzle was no doubt a major reason that "no large amount of timber of merchantable size [has] ever been available."[45]

Betts and Smith did note that eucalyptus "gives a quick, hot fire [and] burns with a bright blaze," although comparisons with other native fuelwoods were needed. Even so, mature trees supplied the best fuel, not the young cord-wood being placed on the market. Blue gum dominated production. But *E. globulus* failed most tests for usable timber. Posts and poles decayed rapidly. Larger ones lasted longer after being treated with creosote, but the extra expense in handling them made the naturally heavy poles more costly than lighter ones fashioned from native redwood or cedar. The best use, concluded the foresters, seemed to be marine pilings. But again, old, slow-grown specimens were preferred, and owing to their weight the piles were "of course more expensive to handle than other timbers."[46]

Betts and Smith's views contradicted McClatchie's upbeat treatment of the timber question. There was too much uncertainty and risk associated with these trees. Certainly, younger, California-grown specimens were unsuited to the range of outlets investors had anticipated, and entrepreneurs had not yet realized (and were unlikely to realize) expected profits. Betts and Smith even admitted that the small quantity of eucalyptus wood that was being used in general construction and "ship work" had been obtained from foreign sources.

Because California wood was still too young, engineers and woodworkers alike were reluctant to sink capital in such an unfamiliar product. "Thoroughly-dry," eucalyptus wood had every quality of a first-class furniture wood, Betts and Smith explained, but they admitted that they knew of no attempt to manufacture furniture on a large scale. The only unqualified claim they made for eucalyptus was for "insulator pins" (which "have been shipped to Canada and the Eastern States," they added eagerly), for certain farm equipment, and for oil derived from the distillation of leaves and twigs. Even then, "the higher grades" of oil were still manufactured in Australia, commented the authors.[47]

Waiting sixty or more years for California plantings to mature into good timber was not the advice entrepreneurs wanted to hear. But that is what they got. It was brought out in a series of Forest Service questions, which the American consul in Melbourne, Australia, posed to the conservator of forests in the state of Victoria. Trees to be used for milled lumber needed time to mature; ideally, a hundred years for river red gum. The only ready profits from anything less than thirty years old came from mine props.[48]

Back in Santa Monica, the scenic rows of blue gum grew fragmented and tattered as city officials permitted homeowners to chop down larger specimens. Their sinuous roots questing after moisture obstructed drains, buckled street pavements, and damaged the foundations of buildings and homes. Residents felled the offending specimens and chopped them into firewood. The market for seedlings that Kinney had noted in that city twenty years earlier had dried up.[49]

In 1918 forester Merritt B. Pratt added his voice to the reappraisal. He calculated that extensive areas of blue gum had been grown for lumber in a short amount of time. However, he noted, "satisfactory lumber can be obtained only from carefully selected large trees." Pratt concluded that the trees were not very old, so although eucalyptus wood was very tough, hard, and strong, useful for hand tools, it did not make superior fence posts due to the rapid rate of decay of the sapwood. Its best use was as a fuel, but by then, oil and gas had displaced eucalyptus wood.[50]

Such sobering assessments devastated investors. Speculators turned away, looking to irrigation to make increasingly useless timberlands suited for field crops. Eucalypts lost out to irrigation, and also to plans from the California Native Plant Society and the Nature Conservancy to remove them from refuges and parks.[51]

By 1950 only three eucalypt species (blue gum, river red gum, and red ironbark) appeared on a list of fifty useful trees from foreign lands. Restricted to a zone where average minimum temperatures were between −2 and −8°C, where subtropical species performed best, authorities characterized the eucalypts as being useful for windbreaks or landscape ornaments, but little else. It was the turn now of other people in other places to continue the de facto reinvention of Australian landscapes overseas. Californians had grown tired of the formula.[52]

More recently, trials with thirty-six eucalypt species held near Concord, twenty miles northeast of San Francisco, reaffirmed expert opinion that the subgenus *Monocalyptus* does not do especially well beyond Australia. Horticulturists determined that the best results came from species native to higher elevations in eastern Australia, and that *E. camaldulensis* and *E. grandis* performed best in terms of growth rates and therefore suited programs for wood pulp.[53]

Innovators such as Ellwood Cooper and especially Abbot Kinney successfully persuaded state and federal agencies and educational institutions to research eucalypts and promote their cultivation. University of California foresters held trials for scores of species, growing them on state property and in experimental plots. Kinney sparked an interest in eucalypts among federal experts, specifically in the Department of Agriculture and State Department. Above all, this entrepreneur predicted a rosy future for his charges. His strategy, like that of his predecessors, was to make eucalypts attractive to nursery-men, who raised them and sold them to the general public as likely to satisfy a range of uses.

Eucalypts in California demonstrate the crucial role of individuals in introducing, distributing, and popularizing these new trees to make them well known and abundant. There was a faddist aspect to this promotion. It is clear that writers and scientists exaggerated the benefits to be anticipated from growing the trees. Many claims were premature and, when tested, failed to live up to expectations. Enthusiasm grew from the 1870s and persisted through the 1890s, when it bordered on hysteria. Investors expected to get rich and trans-form California into a more beautiful and healthful place. Personal profit could coexist with civic virtue. Individuals would make money by growing eucalypts, contribute to the well-being of their neighbors, and satisfy the needs of future generations of Californians. Such support focused upon the economic, envi-ronmental, and medical characteristics of the plant genus, and may have even-tually resulted in approximately 50,000 acres (20,235 ha) being planted with these trees. But as soon as investors began to cut and sell timber; offer aromatic oils, honey, and medicinal products made from eucalypts; or look for mea-surable improvements in local climates and in the health of residents, they came away disappointed.

In one respect this turned out to be a godsend. The tree was not planted on a huge scale or in an organized and comprehensive fashion. Due to the spotty and limited areal extent, the environmental and social problems asso-ciated with large plantations never arose in California, although eucalyptus is a characteristic feature of its suburban landscape today. These issues, however, did arise in Brazil, where the genus formed part of industrial plans to grow the alien trees in huge numbers for the purposes of providing lumber, fuel, and more recently wood pulp.[54]

California's Eucalypts

Eucalypts were planted extensively in California shortly after 1850, and by the end of the nineteenth century almost one hundred species were being grown in both rural and urban areas of the state. Plantings persisted into the first decade or so of the twentieth century, and today eucalypts are still common and well-known trees in that state.

Looking from under a eucalypt umbrella toward San Francisco's downtown and the East Bay skyline.

Campus groves resembled this one at the start of the twentieth century. (McClatchie, *Eucalypts Cultivated in the United States*)

So-called red gum trees line streets in Sacramento, the California capital. (Photo: Andrew Miles)

Red gums ornament William Land Park, Sacramento. (Photo: Andrew Miles)

Sacramento golf courses, too, feature red gums. (Photo: Andrew Miles)

Early view of *E. globulus* (blue gum) on Ellwood Cooper's central coast ranch around the turn of the century. (McClatchie, *Eucalypts Cultivated in the United States*)

Impressive stand of blue gums near Santa Barbara around the turn of the century. (McClatchie, *Eucalypts Cultivated in the United States*)

Eucalypts in East Lake Park, Los Angeles, around the turn of the century.
(McClatchie, *Eucalypts Cultivated in the United States*)

Twenty-four-year-old manna
gum planted as a shade tree in
Pasadena. (McClatchie, *Eucalypts
Cultivated in the United States*)

Initially, *E. globulus* was targeted for producing poles in California. (Betts and Stowell Smith, "Utilization of California Eucalypts")

Blue gum was also recommended for piers and pilings, as in San Francisco. (Betts and Stowell Smith, "Utilization of California Eucalypts")

5 Industrialists and Eucalypts
Take-off in Brazil

> It was Navarro de Andrade, who by his
> tenacity, capacity of work, investigating and
> method and spirit gave to Brazil this great
> amount of knowledge on the precious Aus-
> tralian tree, putting at the disposal of all the
> Brazilians the results of trials, researches and
> observations done on the behaviour of a
> great number of species, according to the
> environment and utilization.
>
> —ARMANDO NAVARRO SAMPAIO,
> translator's note in *The Eucalypt*,
> by Edmundo Navarro de Andrade (2d ed., 1961)

In the final two decades of the nineteenth century, foresters beyond
Europe's colonial reach showed an increasing interest in the cultivation of
eucalypts. At that time (see appendix 2, table 2) about twenty nations boasted
commercial stands grown mainly for local markets. Some experts began to
envisage grander schemes, expanding the areal extent of eucalypts and raising
their status nationally, only to have their hopes dashed when the trees failed to
match expectations. Events in Brazil took a different turn. Through a combi-
nation of circumstances, the efforts of a young agronomist-turned-forester
named Edmundo Navarro de Andrade (1881–1941) established eucalypt plantations on a sustainable footing there. A professional agronomist, Navarro de
Andrade dedicated his career to researching and disseminating the good name
of this Australian plant genus throughout his native Brazil. Inspired by his
knowledge, enthusiasm, and dedication, successors continued to press the
cause of eucalypts. Today, Brazil is the world's largest producer of eucalyptus
wood—largely due to Navarro de Andrade's singular enterprise.

In 1903, after graduating from the College of Agriculture at the University of Coimbra, Portugal, Navarro de Andrade sailed back to Brazil. Antonio Prado, mayor of São Paulo, where Andrade was born, and president of a railroad company (Companhia Paulista de Estradas de Ferro, or Paulista Railroad Company), offered him a job. The young man was appointed director of a forest farm that railroad officials wished to establish at Jundiai, close to São Paulo. They intended to secure a reliable source of wood to fuel steam locomotives and provide posts and fencing along the tracks. Local woodlands were being felled at an alarming rate, presenting a clear and compelling need to grow wood in order to head off shortages. This was the brief that Prado handed Navarro de Andrade.

In December 1903, as newly appointed director of forestry for the railroad company, Edmundo collected as many native and nonnative tree species as he could lay hands on. He sought to determine which of the ninety-five or so species of woody plants he gathered would best furnish fuel, track ties, fence posts, poles, and planks. He later recalled that included in the company scheme at Jundiai were seeds of "some species of Eucalyptus" that he "had brought from Portugal." He planted additional gums "from seeds that I had collected from trees planted in S. Paulo." Some trees in that city stood tall enough to shelter residents from sun and rain.[1]

When Navarro de Andrade began his project, eucalypts had been growing in Brazil for at least fifty and probably seventy-five years. Information in Rio de Janeiro's Jardim Botanico suggests that Pedro I of Brazil planted *E. robusta* and possibly other eucalypts in 1825. There is also speculation that eucalypts arrived even earlier, as the port was a stopover for British ships bound for Australia after 1788.

It seems that *E. globulus,* and one or two additional species, had been planted by about the mid-nineteenth century. In 1865 there is mention of a Colonel Felipe Belbeze de Oliveira Neri, deputy of Rio Grande do Sul, planting seeds mailed from Montevideo, Uruguay, where one or two specimens had been growing for about ten years. Better documentation goes back to January 1869, when France's Society for Acclimatization, with which Prosper Ramel was so closely linked, mailed eucalyptus seeds to Frederico de Albuquerque in Rio Grande do Sul. With its link to this gentleman, the Paris-based promotional group claimed to have made the first successful introduction of eucalypts into Brazil.[2]

Whatever their earlier history, eucalypts drew Navarro de Andrade's attention, and he, like von Mueller, Cooper, and others before him, became their staunch promoter. The railroad scheme was "an industrial enterprise," he claimed, designed to establish forests capable of furnishing products in comparatively short periods "of economical duration." Noting unexpected spurts of growth among the eucalyptus trees, the young forester switched away from native species, which he and his colleagues judged too slow-growing to satisfy "private business." Navarro de Andrade later recalled how he had initially assumed that fine wood came only from slow-growing trees. Later, he explained proudly how his eucalypts had proved him wrong. The Australian trees furnished both fine timber and excellent fuel.[3]

In 1909, with support from railroad executives, who had purchased more land at Rio Claro, Edmundo set out the most promising species in plantations and established demonstration plots for additional trees. The Paulista Railroad Company bought 6,000 acres (2,428 ha) for this forest farm. There Navarro de Andrade grew seventy-two species of eucalyptus and eighty-three species of Brazilian trees, plus a large number of other plants from North America and Europe.

Soon thereafter, with company backing, Navarro de Andrade traveled to the United States on the first of five plant research and collecting excursions. He wanted to see how eucalypts were doing, but after several months he returned disappointed. American forestry colleagues, he surmised, had mistakenly linked eucalypts to land speculation rather than to scientific forestry.

In 1913 the governor of São Paulo commissioned him to journey to England, India, and Australia. In Australia Navarro de Andrade visited Joseph Henry Maiden (1859–1925), consultant botanist to the New South Wales Forest Service and an authority on eucalypts. Warming to the Brazilian forester, twenty years younger than himself, and carrying on von Mueller's tradition of generosity and expansiveness, Maiden handed over seeds from 150 species, plus a similar number of herbarium specimens, which the younger man gratefully carried back to his Rio Claro nursery.[4]

Navarro de Andrade continued the crusade for his favorite trees. During his career he started seventeen forest farms for the Paulista Railroad Company. Between 1918 and 1926 he visited "all the countries in which Eucalyptus is cultivated, even on a small scale," returning from each excursion with seeds for his Brazilian research. In the course of his trials, he discovered that provenance—that is, the point of origin of seeds within the natural range of a species—makes a great deal of difference as to whether a eucalyptus species succeeds or fails in a new location. He noted, for instance, that sugar gum *(E. cladocalyx)*, endemic to South Australia, grew poorly from seeds collected in Australia, but did much better from seeds collected in southern California, where it had done well as a street tree. He also noted that some eucalypts from Western Australia failed in Brazil, but established themselves in Chile, Cape Colony in South Africa, and California—that is, in western locations in each region. Navarro de Andrade grew the majority of the seeds donated by Maiden. He discovered that twenty-five species proved to be excellent, fifty good, and the remainder of little value. It appears that before this infusion of stock directly from Australia, foreign seed originated in France and from two California-based seed firms.[5]

Initially, his colleagues and others did not share the Brazilian forester's enthusiasm for eucalypts. In fact, the public showed hostility and distaste for the genus, declaring that gum trees dried out soils and attracted leaf-cutting ants. It "was received with fire and sword," Navarro de Andrade remembered, "like an undesirable foreigner in whom all the defects were recognized and in whom all virtues were denied." This hostility was ironic, he mused, given the nation's dependence on alien coffee trees and its inability to secure a profitable market for its native rubber plants.

Nonetheless, with Prado's support, Navarro de Andrade pressed ahead. He recorded physiological, environmental, and ecological data about his nursery specimens. He later claimed without exaggeration that in dedicating thirty-eight years of continuous and exhaustive research to eucalypts, and with scrupulous attention to trials, mostly between 1905 and 1915, he had established his botanical progeny on an industrial footing. A downturn in the fortunes of these exotic plants in foreign nations, Navarro de Andrade concluded, was due to the "lack of a scientific foundation . . . the technical character of the work, and chiefly a lack of tenacity and continuity in the studies." This Brazilian took great pride in the painstaking task of broadening the science of eucalyptus in the New World.[6]

Other scientists in Argentina, Uruguay, California, and even the Hawaiian Islands, where large plantations were also started, had not methodically scrutinized the tree's performance or subjected it to empirical tests, declared Navarro de Andrade. Prior to his own commitment, only faddist aspects of eucalyptus cultivation had encouraged propaganda for the trees. Not one to hide his accomplishments, however, Navarro de Andrade claimed with justification to have succeeded when other workers had failed and always to have labored "under the full fire of the enemy's batteries."[7]

Navarro de Andrade concentrated on getting trees to germinate, grow quickly, and look healthy. He proclaimed with unquenchable vigor that "eucalyptus is the only tree yielding first quality wood that can be utilized by the same man who planted it." He noted the tree's many functions, one of which was to free heavy industry from the burden of importing coal from the United Kingdom. He argued that eucalypts provided excellent fuel for steam locomotives, and stated that his employer, the Paulista Railroad Company, had made them generally available in the mid-1920s, thereby saving on outlays for foreign coal. Obligated to fence rights-of-way, railroad officials drew on the company's crop of eucalypts in forest stations to set fences and poles along approximately a thousand miles of track. Navarro de Andrade calculated that the company had supplied 262,517 timbers for its own tracks and those of other railroad companies. Savings on indigenous woods were impressive.[8]

Navarro de Andrade also conducted research on the manufacture of pulp. Initial tests in Australia suggested that the shortness of fiber and density of wood made eucalyptus an inferior source of wood pulp. He contested that opinion, correctly, as it turned out. During a visit to Madison, Wisconsin, in 1925, he experimented with others to produce "several qualities of paper from Eucalyptus" (probably Sydney blue gum *[E. saligna]*, common in southern Queensland and coastal New South Wales). This product was used for an edition of the *Wisconsin State Journal* on 30 December of that year. Back in São Paulo, a paper mill began to manufacture eucalyptus pulp a year or so later, using 25 percent to 30 percent of bleached pulp in printing and writing papers.[9]

At that time, the Paulista Railroad Company owned some 8 million trees on its holdings, mostly intended for fuel and fencing, not paper. Pioneer efforts by Navarro de Andrade encouraged other railroads to set up their own man-made forests. At least four other railroad companies in the state of São Paulo planted 35,800 acres (14,488 ha) in eucalypts for fuel and timber. By 1941 they were growing 10 million trees, and the central government was also investing in nurseries for its tracks.

To facilitate an increase in wood production, lawmakers passed a Forest Code that instructed wood-using industries to replant portions of cutover lands. In a decree of 6 March 1918, state authorities boosted growth by offering subsidies for planting eucalypts and other species. As the ratio of forests to total land area declined, officials used this law to expand the area under tree cover. They recognized that earlier massive clearances for coffee plantations had bitten deeply into native forests.[10]

Navarro de Andrade, roving ambassador for eucalyptus, understood this need to replant with useful woods. He summed up his life's work in a speech he made upon receiving an award from the American Genetic Association in June 1941. B. Y. Morrison, chief of the Section of Foreign Seed and Plant Introduction in the U.S. Department of Agriculture, presented the Meyer Medal, which added to previous international recognition. Portuguese and French dignitaries had earlier honored this Brazilian's distinguished services for plant introduction and research. In this acceptance speech, Navarro de Andrade characterized himself as "a man of action, a forester," who had "spent much of his life alone among Eucalypts, away from the constant chatter of the inhabitants of cities." He had emerged periodically from his "cloister" on the Rio Claro, he said, in order to express an unconditional admiration for his Australian plants. He followed in the footsteps of Ferdinand von Mueller, who, like himself, was a collector, scientist, and true ally of gum trees.[11]

When Navarro de Andrade died, a few months after receiving this recognition, at least 200 million eucalypts had sprouted in his native Brazil. The Paulista Railroad Company owned about 21 million of these and continued planting at a rate of 2.5 million trees per year. Through Navarro de Andrade's zeal, Brazil was then proud possessor of the largest number of eucalypt trees on earth, outside of Australia.

During his management of the Paulista Railroad Company's forest estate, Navarro de Andrade established forest farms along its tracks. He expanded them from 8,000 ha (19,770 acres) in the 1930s to more than 20,000 ha (49,420 acres) at the time of his death in São Paulo. A relation, Armando Navarro Sampaio, took over operations (until 1964) and continued to enlarge eucalypt holdings to approximately 45,000 ha (111,195 acres). The government assumed control of company lands in 1964, but continued to expand the area under Australian trees, which totaled 57,541 ha (142,184 acres) in 1973. That year it sold 5,000 kg (11,025 lb) of seed. Thereafter, the area of company lands under eucalypts declined, as diesel oil and electricity replaced the need for wood. Rio Claro became a historical monument, replete with gardens and lakes, where officials celebrated the centenary of Navarro de Andrade's birth.[12]

Since Navarro de Andrade's day, numbers of eucalypts in Brazil have skyrocketed, as has the area given over to forest plantations. In the 1960s a revised

Forestry Code offered monetary incentives for new man-made forests. A 1966 federal law (5106/66) permitted industrialists a huge rebate on income taxes if they invested in reforestation. Under this law and another one passed in 1970 (Law 1134/70), offering extra-favorable terms, the amount of land under plantations jumped from about 500,000 ha (1.24 million acres) (before incentives) to 6.2 million ha (15.32 million acres) by 1992. In the first ten years about nine thousand applications for replanting were considered, and in the rush hybridized seed produced plantations of uneven quality. After 1970 several companies specializing in reforestation projects emerged. They achieved better results by using fresh, but expensive, imported seed. Over the twenty-one years in which these fiscal incentives operated (1967–88) no fewer than 5.5 million ha (13.6 million acres) of Brazil went under plantations, more than half of which was devoted to eucalyptus.[13]

The data that Navarro de Andrade had generated from his research at Rio Claro furnished valuable guidance to scientific and industrialist successors anxious to promote tree upgrade through grafting, cross-pollination, and tests for disease resistance. Eucalyptus is "the perfect healer" in Latin America, declared forestry consultant C. M. Flinta in 1956. Humans and livestock seek shelter under its branches, he noted, and its upright trunks serve as windbreaks and make valuable fuel. Industrial plantations of such speedy growers replaced native forests as supplies of indigenous woods dwindled.

In the 1950s and 1960s eucalypt stands were springing up elsewhere in South America, though at slower rates. They existed in the interior sections of Argentina, Uruguay, and Ecuador. Riparian areas of the rivers Uruguay, Paraná, and Paraguay also supported woodlots composed of five species and reportedly valued as pulpwood, fuel, and timber (for fruit crates). Eucalyptus trees ornamented resorts and anchored large sand dunes along the Atlantic coast. In the far west, they grew splendidly. In Peru's altiplano, where the trees had existed for a hundred years, industrialists turned to blue gum for fuelwood, pit props for mines, and railroad track ties. In Chile, probably the first country in South America to grow eucalypts, blue gum found similar uses among indigenous people as well as among German immigrants in central and southern regions. Eucalypts apparently found a home in Argentina directly from Australia, but arrived in Ecuador via France. Farther north, in Colombia, for example, the first seeds came from the United States. There, the genus has not yet grown as popular or prospered as well as it has farther south.

In the 1950s there was also a basic need for forest scientists to conduct trials, like those of Navarro de Andrade in São Paulo. Growers needed to exchange information and instruct colleagues at home and abroad about what and where to plant. At that time experts began to confirm that hybridization had taken place between eucalypts planted together (as at Rio Claro) and to recognize that adequate provenance trials, especially when duplicated, expanded, and subjected to the kinds of analysis that typified Navarro de Andrade's approach, would enhance industrial development.

The dearth of technical assistance and the need for government agencies to improve technical skills in land-use planning, forestry, and horticulture inhibited eucalypts from becoming more widely used. However, the Second World Eucalyptus Conference, which convened in São Paulo in 1961, helped rectify this situation.[14]

Participants at this conference expressed "highest consideration for the work undertaken at Rio Claro by Navarro de Andrade and his successors."[15] At that time, Brazil had some 560,000 ha (1.38 million acres) in eucalyptus plantations, much of it devoted to blue gum, lemon-scented gum, forest red gum, and other, less important species. About 80 percent of that nation's plantations were located in the state of São Paulo, blanketing about 2 percent of its land surface. Growth densities averaged 2,000 trees per ha (810 trees per acre), and nursery output totaled 270 million seedlings annually. Brazil's eucalypts represented about 80 percent of all those in Latin America. Most went for low-sulfur charcoal in the manufacture of pig iron and steel.[16]

Delegates at the 1961 conference applauded breakthroughs in growth and utility. An Argentinean research team and the company Celulosa Argentina, based in Buenos Aires, led the charge. Chile had its Sociedad Agricola, interested in gum trees, and in Brazil the Servicio Forestal of the Companhia Paulista based at Rio Claro continued Navarro de Andrade's pioneering research and development.[17]

By the early 1960s the switch to plantations was paying off. A surface area of 800,000 ha (2 million acres) given over to eucalypts in Latin America was double that in the Mediterranean basin, and more than double that in Africa. The standing crop was, of course, far less than that in Australia, but conference delegates learned that growth rates in Latin America were proving to be actually higher than in many native habitats in Australia.

Expectations soared, especially after Italian-born forester Lamberto Golfari pressed ahead with so called "exotic forestry." With backing from the Food and Agriculture Organization of the United Nations (FAO), Golfari dispensed basic ecological data among growers, seed merchants, and wood producers in Brazil. He divided the country into twenty-six bioclimatic provinces and determined what eucalypts grew best in each of them. For example, Golfari recommended *E. urophylla* and two additonal species for São Paulo's inland location, and *E. saligna* and *E. grandis* for moister soils along the coast. Interestingly, the suggested provenance for *E. grandis* seeds was Coffs Harbour, New South Wales, Australia, where paper interests had planted eucalyptus on degraded farmland in the late 1950s. Scientists discovered a decade later that flooded gum grows rapidly, is quite adaptable, and well suited for pulp. Consultant Lindsay Pryor, a professor of Botany at the Australian National University in Canberra, conducted early experiments with Coffs Harbour flooded gums and recommended them to colleagues like Golfari in Brazil. *Eucalyptus grandis* has helped upgrade plantation quality in that nation.[18]

As a consequence of legal and fiscal incentives, by the mid-1970s eucalyptus plantations covered 1.4 million ha (3.46 million acres) in Brazil alone, mostly in Minas Gerais and São Paulo, and plantings exceeded 37,500 ha (92,662 acres) annually. A decade later, that figure topped 100,000 ha (247,100 acres) per year, mostly going for wood pulp and charcoal. However, much of the seed continued to be of inferior quality. Industrialists selected poor sites, and as a result harvests continued to be unprofitable. Visiting Australian forester Douglas Boland reported that only one hundred of about twelve hundred forest companies were economically significant, and perhaps ten of these merited international attention.

Brazil has continued to improve eucalypt stocks. It is the largest single producer of eucalyptus pulp in the world and is capturing an ever-increasing share of the annual global supply. Most recent estimates suggest that Brazil has 3.62 million ha (8.95 million acres) under eucalypt plantations (see appendix 2, table 1), making it second only to India.[19]

Eucalypts for Pulp

The cultivation of Eucalyptus is very much
profitable and may thus fulfill a country's
need of raw material, while being at the same
time an interesting field of investment.

— Chilean forestry expert at
Second World Eucalyptus Conference, 1961

Whether we know it or not, eucalyptus wood is part of our daily lives. Eucalypt cellulose is processed into clothing, including undergarments, swimwear, and even military uniforms—fire-retarding and combat-ready. More and more paper household products—including bathroom tissue, wrapping paper, cardboard, and specialty bond papers— come from eucalyptus logs that have been chopped and emulsified into pulp.

In the pulping process, felled trees are hauled, cut into chips, then fed into digesters boiling with chemicals that liberate natural wood fibers, including a dark by-product called lignin. The resulting pulp is washed to remove the chemicals, bleached with chlorine dioxide, and rolled into sheets for paper or spun into viscose for furnishings and garments. Other uses for the pulp include hardboard, a range of high-quality tissues, and packaging. Water, wood residues such as lignin, and reductive chemicals are siphoned off and heated in a recovery boiler that turns the water into steam and wood residue into fuel; the chemicals are then purified for reuse. Recycling solvents is essential in modern pulp production, while timber derivatives, such as lignin, hemicellulose, and even bark and twigs, fuel driers and generate electricity for mill operations.[1]

Although softwood pulp fibers are longer than hardwood fibers and make stronger paper, the shorter hardwood fibers, exemplified by eucalyptus, make a better surface for printing. They are fairly strong, bulky, opaque, and easily dewatered and dried. In the United Kingdom, for example, a typical sheet of paper consists of 55 percent hardwood pulp, mostly from eucalyptus, 20 percent softwood pulp, and 15 percent filler (chalk or special clays).[2]

Tests and trials in the manufacture of eucalyptus pulp began more than eighty years ago in Australia and in Portugal reportedly go back to 1906. In Brazil, as we have seen, Navarro de Andrade also experimented with pulp making. In 1927 the firm of Gordinho Braun, the first to manufacture sulfite pulp in Brazil, used its own bleached supplies from eucalyptus, pioneering this chemical pulp-making process in South America. Using acid chemicals at high temperatures, sulfite pulping breaks down wood fibers and dissolves and removes lignin. Pulp yields, however, are lower than 50 percent of the total weight of wood. The end product is much stronger than pulps derived from mechanical processing alone, which grinds up wood chips in the presence of water. Mostly softwoods, and not short-fibered hardwoods, such as eucalyptus, are processed by sulfite and mechanical pulping. It was not until soda and semi-chemical processes loosened the bindings of these stiff fibers, while conserving their strength, that eucalypt pulp began to be included in papermaking.

In the 1950s the chemical "cold soda" process was successful in pulping *E. saligna* grown on Paulista Railroad Company holdings. A paper factory in Canada and another in the United States used this eucalypt to supply 20 tons (18.2 metric tons) of newsprint, which was used for an edition of *O Estado de São Paulo* that appeared on 27 May 1956.

The 1961 Second World Eucalyptus Conference in São Paulo explored advances in pulp making and papermaking by using eucalypts. One speaker explained how eucalyptus could go into newsprint and magazines. In the cold soda and cold soda semi-chemical pulp operation, wood is impregnated with sodium hydroxide at room temperature, then mechanically processed to obtain the defibration of the wood chips. The resulting pulp is coarse, cheap to make, and of low strength; it may be mixed with long-fibered pulps and processed to make papers and paperboard of varying strengths and brightnesses. The process is unsuitable for softwoods. Cold soda pulps from eucalypts are used as components in newsprint, fine paper, and magazines.[3]

Conferees in 1961 also discussed alkaline pulping using sulfates. In this so-called kraft (from the German word for "strong") pulp treatment, wood is steamed under pressure in a digester filled with sodium hydroxide and sodium sulfide. Crude sulfate pulp makes a dark-brown pulp suited for paper bags, or it can be bleached to high brightness for printing and writing papers. This sulfate (kraft) process turns out high-strength pulp suited for most grades of paper and is in use throughout the world. Improvements in chemi-mechanical and thermo-mechanical pulping raised yields in terms of the weight of wood used (from 60 percent to 85 percent of pulp generated from amount of wood pressed) and helped establish a range of outlets for eucalypts in the 1970s.[4]

The usefulness of eucalypts for pulp depends upon what species is used, where it is grown, and what it is intended for. Blue gum pulp, for example, has a low to average length and width among hardwoods and is thin-walled. This gives it flexibility and a good fiber for bonding in finished paper. Blue gum is strong, bleaches to high levels of brightness, takes cold soda processing well, and can be processed without removing the bark. Wood from young, fast-grown river red gum (E. camaldulensis), the second most widely grown plantation species, is very similar, generating pulps of good opacity and excellent strength.

Forest red gum (E. tereticornis) consists of fibers of average length and thickness for hardwoods. It too delivers superior opacity and brightness. But this tree is difficult to debark, and in sulfate pulping, in which wood is digested in a heated bath of sodium chemicals, yields of crude pulp are low. It is also expensive to bleach.

Over the last thirty years, additional eucalypts have entered the pulp markets. One of them, lemon-scented gum (E. citriodora), which grows well in Latin America and Africa, also tends to be difficult to debark and makes into a rather stiff paper. Kamerere (E. deglupta), the eucalypt native to the Philippines and New Guinea, is grown on Pacific islands, in Cuba, and in parts of India and Brazil. But its pulp is not especially good. Another one, however, mountain ash (E. regnans), does render high-quality material. It furnishes unbleached pulp of excellent strength and brightness, the sort needed for newsprint and writing paper. Its low-density, straight-grained wood is included in all types of mechanical and chemical pulping procedures, and inclusion of the bark increases pulp output.[5]

The South African provinces of Natal and Zululand grow large expanses of flooded gum (E. grandis, also called rose gum), which shoots upward very swiftly. In recent years a large factory near Durban has been turning out 450,000 tons (457,530 metric tons) of rayon-grade dissolved pulp for both the fiber and paper industries. Flooded gum has become the tree of choice in Brazil. Its light-colored, softer wood is strong and durable. Thin wood fibers are strong yet possess a flexibility that makes them bond together in paper products. The tree's high lignin content, however, poses problems in the treatment of pulp effluents.[6]

Currently, chemical pulps made up from eucalyptus are excellent for the manufacture of fine papers, because their fibers are short, thin, and flexible in comparison with other hardwoods. The large numbers of fibers per unit of mass gives high opacity; stiffness adds bulk and strength. These useful properties need to be safeguarded as foresters continue to improve both the growth rates and the genetic composition of gum plantations.[7]

Scattered throughout Africa, Asia, and Latin America, pulp-producing factories are increasing production in order to meet both regional and world demands for paper and paper products. With world paper consumption expected to rise, from about 270 million metric tons in 1994 to more than 480 million metric tons by 2010, world pulpwood consumption will also grow. And pulp mills in Brazil will remain major suppliers.[8]

One of the largest "eucapulp" operations is Brazil's Aracruz Celulose S.A. (with its subsidiary Aracruz Florestal in charge of plantations). Aracruz is a good example of the size and scale of wood pulp operations. From Navarro de Andrade's solitary, steadfast quest to gain acceptance for eucalypts, a behemoth has emerged that generates vital foreign currency for the Brazilian economy. International financiers have bucked that nation's recent inflationary trends by investing in the burgeoning market for eucalyptus pulp.

Aracruz Celulose S.A.

Millions of eucalypts surround the massive industrial complex of Aracruz Celulose, located on the coast of southeast Brazil in the state of Espirito Santo. The enormous pulp-making factory is 400 km (250 mi) north and east of Brazil's major city of Rio de Janeiro. It is a wood pulp factory consisting of two mills, a chemical plant, a town with more than five thousand workers, and plantations mostly growing eucalypt clones, arranged in mosaics of one hundred hectare-sized blocks. There is a port complex through which pulp is shipped principally to Europe and North America. Administrative headquarters are in Rio de Janeiro.

Aracruz is the world's largest producer of bleached eucalyptus pulp. It manufactures about 20 percent of the world's output of eucalypt kraft market pulp—that is, about 3 percent of overall world production of wood pulp. Over the past decade the factory has dominated Brazil's pulp industry, accounting for approximately 50 percent of that nation's eucalypt pulp exports and assuring Brazil a dominant role in the export of bleached, short-fiber pulp. Aracruz is one of Brazil's five major pulp-and-paper firms, which are grouped into the Brazilian Pulp Exporters Association (ABECEL). It is the first paper-and-pulp producer in that nation to be listed on the New York Stock Exchange (see appendix 2, table 3).[9]

Paradoxically, eucalyptus pulp production in eastern Brazil got its real boost from four centuries of literally shredding native forests along the Atlantic seaboard. Despite penalties aimed at prohibiting deforestation and the burning of native forests, settlers turned aboriginal woodlands into fuel and lumber. Farmers cut and cleared trees, including valuable Brazil-wood, and turned the new spaces into pastures and plantations for coffee and cacao. In 1831 all laws limiting tree harvesting and burning were abolished. Today, only a tiny fraction of native woodland remains from what authorities estimated was once a swath as much as 200 km (125 mi) wide along the Atlantic seaboard. Scrub, pasture, crops, and urban expansion have bitten deeply into aboriginal tree cover, and surviving fragments of it continue to be lost. Over the last thirty years, however, larger and larger tracts have been revegetated—not with the more than fifty indigenous trees that foresters esteemed, but with green-crowned plantations of look-alike eucalypts.[10]

The decision to establish and expand areas under Australian eucalypts goes back to information gained from the industrial trials of foresters like Navarro de Andrade. Tree experts studied the commercial viability of the genus and were impressed. Government subsidies to revegetate portions of the state of São Paulo and elsewhere through low-interest-rate loans and tax exemptions made the cultivation of eucalypt plantations a winning proposition in the mid-1960s. Up to half a firm's tax liability could be written off when it invested in planting trees. A second law allowed tax reductions of as much as 20 percent for tree planting. Both incentives resulted in a major expansion of the area under eucalypts.[11]

The Eucalyptus

Anticipated international shortfalls in the production of pulp for paper-making in the 1960s contributed to the multiplication and spread of plantations. Experts estimated that no fewer than 150 new pulp mills costing some $10 billion were needed to head off worldwide paper shortages. Brazil, Chile, and Argentina were touted as the best growers and manufacturers.[12]

With government incentives lasting into the mid-1980s, eucalyptus growing boomed. Between 1966 and 1973 almost 300,000 ha (741,300 acres) of land went into gum trees, mostly intended as fuel for the steel industry in the state of Minas Gerais. Some of it was sold at hugely inflated prices. As land values increased, local people sold off poor-quality holdings for plantations and expected that corporate buyers would employ them. In Espirito Santo, Aracruz Celulose holdings began to contribute significantly to Brazil's eucapulp industry as soon as its massive processing plant was finished (see appendix 2, table 3). The industrial complex consisted of chemical and milling units that covered a strip of land 1.5 km long that ended in port and harbor facilities. In 1979, after the first year of operations, the complex manufactured 291,000 metric tons of pulp. It sold 286,000 metric tons that year and generated exports in excess of $100 million.

In its first five years of operation Aracruz earned Brazil $650 million in foreign exchange and attracted international acclaim as a major world supplier of bleached, kraft eucalyptus pulp. By its fifth year, Aracruz accounted for 40 percent of Brazil's exports of this product, and it has continued this hegemony.[13]

Plantations

Over the past thirty years, Aracruz has increased its land holdings to 213,000 ha (526,323 acres) in Espirito Santo and Bahia, of which 138,000 ha (340,998 acres) is under plantation eucalypts. Recently, the company has actively encouraged local farmers to grow eucalypts for household purposes, such as firewood, and for selling back to the company for pulp. Every year, company foresters and local farmers set out millions of seedlings in geometric grids on and around the industrial complex. Currently, workers fell 53 ha (131 acres) per day of tall, straight-trunked trees, and huge trucks carry the logs back to the mills. The vehicles make about 275 deliveries every day.[14]

When the factory started up, company foresters planted 59,000 ha (145,789 acres) with approximately 1 million eucalypts belonging to more than fifty species. Their aim was to grow the best trees, cut them, encourage re-sprouting, then figure out ways to speed up the growing cycle. First cuttings in 1979 came from the original plantings of 1967. Today, that initial twelve-year cycle is down to seven, and the amount of annual growth per hectare of trees has increased substantially, as has the quantity of pulp obtained from the cut wood. Research and development has made eucalypts a tree crop, inten-sively grown, designed and selected to give high-quality cellulose in the short-est time.[15]

Two factors have driven this intensification. First, company personnel selected eucalypts that grow rapidly and furnish superior pulp in Aracruz's warm, moist coastal location. For example, in 1973 silviculturist and eucalypt researcher Edgard Campinhos traveled extensively in Australia, Timor, and Indonesia to secure seeds of *E. urophylla*. The second factor involved research, begun in the mid-1970s with Australian and French assistance, to reproduce quality stock from cuttings rather than seeds. Biotechnicians aimed at upgrad-ing vegetatively derived clones. They selected clones for fast growth, resistance to disease, and pulp quality. Today, company personnel claim that Aracruz's pulp can be custom-grown for domestic and foreign markets.

Two important species, *E. grandis* (flooded gum) and *E. urophylla* (Timor mountain gum), and their hybrids, have figured into this research (see appen-dix 1). The two species are speedy growers and make excellent pulps. Flooded gum grows in northern New South Wales and Queensland, Australia. Histor-ically, *E. grandis* and *E. saligna* (Sydney blue gum, which also gives excellent pulp), have been confused taxonomically, due to a *saligna* designation given to *E. grandis* specimens. In 1918 Australia's botanical expert Joseph Maiden reex-amined and separated the two species. By then, many plantations in foreign countries, including Brazil, were called "saligna plantations" but probably con-sisted of *E. grandis* specimens. The flooded gum has a straight, smooth, white or silvery trunk whose bark can be removed more easily than that of *E. saligna,* whose trunk has a bluish or greenish tint. Ease of debarking, plus a very rapid rate of growth, in excess of 3 m (10 ft) per year for the first decade, have made flooded gum popular among pulp manufacturers in Brazil.[16]

E. grandis is also made into charcoal used to smelt iron. As a wood, it is easy to saw, machine-finish, and nail. This species also hybridizes with other gums. In the first years it is susceptible to termite attack and canker disease, but recently clones have reduced this threat on Aracruz plantations.[17]

E. urophylla is native to Timor and other Indonesian islands. Typically, it grows closer to the equator than the flooded gum and occurs at altitudes up to 3,000 m (9,800 ft) above sea level in its native habitat. Yet this pulp-bearing tree has proven very adaptable in Brazil, where Navarro de Andrade introduced seed from Java into his arboretum at Rio Claro. He laid out three plots consisting of 400 trees, which in turn became the source for innumerable plantings.

The species was often referred to as "Brazil alba," suggesting a misnomer between this species and *E. alba*. *E. urophylla* and the hybrids cultivated by pulp processors grow rapidly in moist, warm lowlands. Aracruz is a good site for it, and it appears to resist disease, such as canker, better than *E. grandis,* although, as was the case for *E. grandis,* termites infested the first plantations.[18]

In 1984 a Swedish organization, the Marcus Wallenberg Foundation, awarded the four-member Aracruz forestry team, including Edgard Campinhos, a forestry prize for advances "in the development of large-scale eucalyptus plantations by means of the cloning method of vegetative propagation." Aracruz scientists had experimented with cloning for about ten years before they gained international recognition and found that their work was beginning to pay off in economic dividends. By 1989 virtually all the company plantations consisted of eucalypt clones. Asexual reproduction through cuttings has improved nutritional uptake, increased wood biomass, and improved overall resistance to disease. Planting clones also facilitates adjustments to site peculiarities and simplifies processing methods to ensure the even quality of pulps.[19]

In order to grow these strains, researchers have had to pay close attention to the biophysical characteristics of plantation sites around mill complexes. Their aim has been to raise wood intensively, while including native plants as part of the plantation ecosystem. They claim that interplanting on a ratio of 2.4 ha (5.93 acres) of eucalypts to 1.0 ha (2.47 acres) of native woodland helps keep man-made forests free of disease. Interplanting also reportedly improves productivity in selected situations; for example, *E. grandis* yields now average 55 m³/ha (22.3 m³/acre) annually. Between 1976 and 1993, Aracruz's eucalypts averaged 29 m³/ha (11.7 m³/acre) per year, increasing to 44.6 m³/ha (18 m³/acre) for five years ending in 1991. Recent reports from southeast Brazil claim additional gains, exceeding 100 m³/ha (40.5 m³/acre) on experimental plots.[20]

Marketing Pulp

Eucalypt pulps make good specialty papers and household tissues due to their strength, density, high opacity, and softness. From the 1950s through the early 1960s, manufacturers used them as fillers in the production of paper and newsprint, mixing shorter with longer softwood fibers. Today, this situation is very different. In Brazil, eucalypts are a virgin pulp, capable of being used exclusively for tissues and high-quality writing papers. Mill owners pride themselves on being able to make a pulp to fit a client's needs.[21]

Over the past thirty years markets for eucapulp have made sizable gains. Hardwood eucalyptus has been a major growth sector in the paper-and-pulp industry worldwide, growing more rapidly than other kinds of pulp due to its versatility and low production costs. Aracruz's output reflects this, as production records were smashed over and over again in the 1980s and early 1990s. The pulp market, however, must also respond to "adjustments," which means that there are good times and bad, depending on the world economy and specifically on pulp sales to Japan, the United States, and western Europe.

In the first five years of operation, Aracruz executives decided to make the plant self-sufficient in regard to fuel and chemicals and hence boosted pulp production and sales (see appendix 2, table 3). By 1984, using new debarking machines, officials had increased plant capacity to 456,000 metric tons per year (from 361,000 in 1980) and were using tree bark as a fuel in the plant. In 1984 exports generated $158 million in foreign exchange. Most of the 439,000 metric tons sold that year found outlets in the United States, United Kingdom, Belgium, and West Germany. Confidence was high, and debts on the mill began to be paid off at a brisk pace.

In 1987 company executives decided to double the capacity of Aracruz Celulose. They designed a new mill that would increase output to more than 1 million metric tons of pulp per year. This new facility, called Mill B (see appendix 2, table 3), began operations in 1991, just as a major recession hit the industry. Although worldwide capacity for bleached, short-fiber market kraft pulp grew 7 percent (to 12 million metric tons) that year, mills expanded production and created a surplus of pulp that weakened the market. In Aracruz, production averaged 89 percent of overall capacity, and with a decline in pulp prices by about $170 per ton (from $708 per ton in 1990), this Brazilian producer and others braced for financial hardship.

The depressed pulp market lasted two years. In 1993 Aracruz cut its plant work force by almost a quarter and reduced pulp output. But it still managed to sell more than 1 million tons, although at a record low price of less than $350 per ton. Then company officials decided to buck the market and increase pulp production. They opened elemental chlorine free (ECF) and totally chlorine free (TCF) pulp lines. These alternative bleaching techniques lower the toxicity of effluent in the milling process—an issue among consumers and environmental interests. Strategies to both treat and recycle such chemicals used in the manufacture of pulp and to "close the loop" by using less and less water in manufacturing are more friendly toward the environment. Ultimately, such techniques make economic sense, too, by reducing financial outlays for pulp processing. Aracruz is an important player in the trend toward adoption of bleaching processes that phase out conventional uses of chlorine and its compounds. In 1989 Aracruz began to make the switch from molecular chlorine, used in traditional kraft hardwood bleaching, to ECF pulp. Two years later, with the second mill operating, engineers took additional steps to abolish chlorine and dioxin residues. The new TCF pulp is an excellent product whose process-related residuals are most easily absorbed by surrounding air and water.[22]

This revamping of pulp production has been achieved without increasing production costs or degrading pulp quality. Aracruz prides itself on being one of the lowest-cost pulp makers in the world, and costs per ton of pulp are significantly lower in Brazil than in other eucapulp-producing nations. In 1988, for example, the cost of manufacturing bleached short-fiber kraft pulp in Brazil averaged $392 per ton, compared with $453 per ton in Portugal and $528 per ton in Spain. These two major competitors also have a longer plantation cycle for eucalypts (between eight and twelve years), and productivity per unit area is lower.[23]

Trends

In June 1995 Luiz Kaufmann, chief executive officer of Aracruz Celulose, declared that he was seeking new outlets for wood products. Diversification was the watchword, he noted, because although demand and prices for pulp had risen (to $875 per ton), shifting Aracruz from loss to profit, the volatility of the pulp market disturbed him. It was time to decide whether eucalypts could be more than pulp—for example, whether the trees could be used for construction and furniture timbers, including compounds for plywood and fiberboard. Success with clones makes this a distinct possibility. A few species, maybe ten of the two hundred or so tested by Aracruz, are suited for other uses, Kaufmann noted.[24]

Diversification does not mean that the industrial complex will decrease production of pulp. Today, Aracruz accounts for almost half of the hardwood sulfate pulp manufactured, mostly from eucalypts, in Brazil. And Brazil, in turn, accounts for at least 60 percent of South America's pulp-and-paper supply (with Chile a distant second with about 10 percent). It does mean, however, that eucalypts may finally reach the range of outlets that pioneers predicted for them, and that eucalypts grown in other parts of the world will prove as useful as any in their native Australia. With new markets in mind, experts expect overall output of fiber to increase by 13 percent by the year 2000. Analysts predict that $10 billion will be invested in Brazil's pulp-and-paper industry before the year 2005. Currently 215 companies employ more than 100,000 workers in the manufacture of newsprint, paper, and tissues, with a combined value of 2 percent of the nation's GDP. Efforts to upgrade pulp lines, reduce production costs, and install advanced technology, as Aracruz has done, are being replicated by other firms. Brazilian industrialists, who rank among the world's most efficient and productive producers of pulp and paper, now grow softwood pines twice as fast as their counterparts in Europe or North America are able to do, and they grow hardwood eucalypts on a six-year cutting schedule. Eucalypts cover 857,000 ha (2.1 million acres) of 1.5 million ha (3.7 million acres) of recently replanted land. The experts' optimism stems from calculations that the world demand for all wood products will grow by two-thirds in the first twenty-five years of the twenty-first century (to 6.6 billion m³/yr). Demand is best met in their scenario by forest plantations. Fast-growing trees, including eucalypts, could meet present global wood demand from merely 150 million ha (370 million acres; less than 1 percent of the earth surface). Therefore, Brazil's combination of climate, soils, cheap land, and low labor costs, and its track record with planting and upgrading plantations make it a natural leader in applied forestry and wood output.[25]

After record pulp prices as high as $820 per ton in the early 1990s, world prices dropped to $520 per ton and have fluctuated due to economic downturns in both Brazil and Asia. However, investors are bullish about eucapulp in Brazil. Sales revenues top the $1 billion mark annually, and exports of bleached short-fiber pulp increased 5 percent between 1996 and 1997. Currently, annual per-capita consumption of paper in Brazil stands at 27 kg (60 lb) versus 298 kg (657 lb) in North America. Paper consumption in Brazil is growing, as it will in other developed nations, so it is clear that eucalyptus plantations and mills for processing the trees will need to expand or multiply in order to satisfy new demands worldwide.

Currently, pulp-paper establishments turn out at least 9 million metric tons of eucalyptus cellulose every year—only 50,000 metric tons was produced worldwide in 1960. Today, Aracruz alone mills twenty times that total. Brazil's dominance is beginning to be challenged, however, by manufacturers in Chile, where pulp makers are switching to eucalypts. Some Chilean companies have already abandoned long-fiber pine softwood pulp, because conifers take double the time that gums take to reach the desired size for cutting.[26]

In Chile, the firm Forestal Colcura pioneered blue gum logs as mine props, then grew the same species for pulp, judging it to be flexible in regard to site. Expansion by the nation's pioneer of bleached eucalypt pulp, Forestal e Industrial Santa Fe, toward 1 million tons per year shortly after 2000 will expand the Grupo Santa Fe's eucalypt plantation area to 55,000 ha (135,900 acres) and intensify production through cloning techniques. Eucalypt forests are also taking hold in Argentina, Uruguay, Ecuador, and Peru.[27]

Another major player in the scramble to increase per capita consumption of paper both at home and abroad is Indonesia, which has four huge pulp mills generating bleached hardwood pulp from eucalypts. The plants are located on East Kalimantan, Sumatra, Jambi, and East Java; recent economic recession has affected the process of expansion.[28]

Environmental Concerns

An increase in the number and size of eucalypt plantations, and the pace at which foresters plant them, have attracted scrutiny from both public and private sectors. In 1993 Aracruz Celulose, which claims a progressive stance toward sustainable development, was ordered by a Brazilian judge to suspend all plantings (then running at millions of eucalypts per year) until further notice. The ruling came in response to a suit filed by the Brazilian Environmental Agency and the state of Espirito Santo. These agencies and other groups accused Aracruz and other pulp manufacturers of felling indigenous forests in order to make room for eucalypts. Such exploitation, the plaintiffs argued, contravened a 1990 federal law. The activist environmental organization Greenpeace had filmed workers from a pulp company illegally chopping down native woodlands in a section of the Atlantic coast plain of Espirito Santo and Bahia—an area where indigenous vegetation had already been reduced by more than 90 percent.[29]

Foresters, environmentalists, and others note that plantations of even-aged, quick-growing trees contribute to loss of biological diversity when compared with indigenous forests. Logging operations also increase soil erosion. Air pollution results from pulp processing, and mill effluents pollute coastal fisheries. Claims have been made of appropriation of native-owned lands; for example, the 1993 suit referred to lands held by the Tupiniquim Indians that had been appropriated for eucalypts. Critics also accuse Aracruz of contracting with farmers to plant the Australian trees in order to circumvent a company pledge to restrict plantations.[30] On this last point, company officials explained that they had donated seedlings to local farmers to provide lumber and charcoal in a region long denuded of its indigenous vegetation. Rather than being damned for creating a "green desert," company foresters argued, they should be praised for ensuring that eucalypts would spring up in a mosaic of patches shared with native plants. To substantiate their commitment to local ecology, Aracruz officials pointed out that a company nursery was also growing and planting indigenous species. Moreover, they noted, reserves with indigenous species cover 56,000 ha (138,376 acres)—that is, 27 percent—of the company's holdings on the Atlantic coast.[31]

Effluent standards are also improving, industry executives claim. This is due to the phasing out of the chlorine used in the bleaching process. European and American customers demand ECF and TCF pulps, whose production is less damaging to the environment. Although there are regional differences in pulp mill standards, the Europeans are committed to TCF technology, while Americans favor ECF. But both technologies are part of the trend toward upgrading pulp mills to internationally accepted standards for water and air quality.[32]

Forester Ricardo Carrere, coordinator of the World Rainforest Movement and a critic of industrial forest plantations, accuses Aracruz and other pulp conglomerates of attempting to "greenwash" their activities—that is, attempting to make them appear environmentally and socially responsive to the needs of indigenous landscapes and their residents (plant, animal, and human). "The chorus of praise" given to Aracruz for progressive policies "has little factual foundation," he argues, adding his voice to growing criticism of the entire pulp industry. Carrere criticizes Aracruz's hold on its region, claiming that it expelled thousands of local residents in the 1960s. He also condemns the company's employment practices, claiming that it replaces aging workers with younger workers, equally poorly paid. Around its plantations streams have run dry, and land given back to native people has not been as productive as it once was. Industrial effluents pose health problems for human beings and also harm wildlife. Aracruz, says Carrere, has obfuscated the environmental and social costs of doing business.[33]

There appears to be the beginning of a trend in Latin America toward stronger laws and the willingness to enforce them. About 1,500 Tupiniquim and Guarani Indians have set in motion an international campaign to win back lands planted in eucalypts by Aracruz. In March 1998 they invaded plantations in Aracruz and reportedly destroyed 4 km (2.5 mi) of eucalypt plantings. Since authorities awarded them 2,500 ha (6,177 acres) of their total land claim of 13,500 ha (33,360 acres) in north Espirito Santo, the situation has grown even more tense. Nations like Brazil that are generally noted for entrepreneurial excesses and laxity toward the environment are now beginning to enforce controls. Laws are being tightened in Chile as well, where it is now an offense to cut down native trees. That law, however, did not prevent companies from cashing in on $100 million worth of wood chips slashed from native vegetation in 1993.[34]

Future Markets

In the mid-1990s Brazil's pulp exporters adopted a wait-and-see policy in respect to plans to further expand forest plantations. Experts anticipated a slowing in the upward trend in pulp growth. The heady 1980s had been overshadowed by a world recession. However, eucalypts now flourished in places where it had never seemed possible that they would grow and on scales never before envisaged. Recent estimates suggest that Brazil now has 3.6 million ha (8.9 million acres) of eucalyptus plantations, making it second only to India as a world producer of eucapulp. Brazil's vast plantations, with 1,000–2,000 trees per ha (400–800 trees per acre), make up approximately 25 percent of the earth's eucalypt forests outside Australia.[35]

According to forestry expert Julian Evans, tree plantations in the tropics and hot subtropics extend over some 42.7 million ha (105.5 million acres). He estimates that eucalypts make up 37.5 percent of that total, or 11.4 million ha

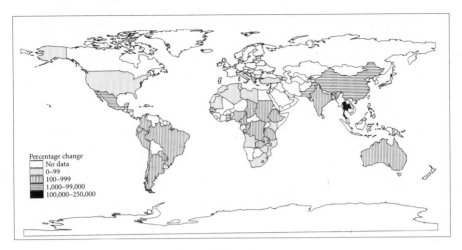

Percentage change of planted area, 1970s to 1990s

(28.2 million acres). A higher estimate has been proposed by FAO consultant John Davidson, who counts 13.4 million ha (33.1 million acres) of eucalypt plantations. Whichever is correct, there is no doubt that a demand for short-fiber eucalypt pulps will persist well into the next century. As a tree crop, the genus is destined to keep expanding its current geographic range, which has doubled each decade since 1960 (see map).[36]

Foresters and industrialists see a bright future for the genus, not just in South America. Stora, a Swedish forest products group, has invested in a Portuguese eucalyptus mill that already has 45,000 ha (111,200 acres) in wood reserves. In this instance, the investor reckoned that eucalypts supply 1 metric ton of pulp from 3 m³ (106 ft³) of wood, whereas birch trees in Sweden supply 1 metric ton from 4 m³ (140 ft³). More important, eucalypts generate this kind of return after only fifteen years, whereas birch takes a minimum of sixty years. British investors also purchased eucalyptus-based mills in both Spain and Portugal in the late 1980s, mindful of the new pulp demands from the European Union. Cropping the trees, rather than "mining" them, supplies the justification for this outlook. It is a position that those opposed to eucalypts find hard to combat. The trees, which can be planted closely together, realize high returns of wood, thereby releasing pressures on native woodlands, especially rainforest hardwoods, about which there is so much public concern.[37] The policy in the pulp-and-paper industry has been to switch from mining trees to renewing them with such plantations. In this approach, trees are a crop to be cut on a rotational basis. Eucalyptus are one of the industry's new components, especially in South America, South Africa, and Iberia. In the cropping approach, workers cut the trees, watch them sprout again from coppice, then cut them again, allowing regeneration two or three more times before they need to plant seedlings. In this way, they establish a thirty-five- to forty-year growing and cutting cycle.

Recent alarms about the environmental and social costs and inequities associated with increased pulp production in Brazil, as well as the growing fears by environmental organizations about the general expansion of plantations throughout the tropical world, can best be explored by turning to the record of eucalypts in India. Soils on that Asian subcontinent have supported expanding plantations of this alien plant genus for a little more than two hundred years. Therefore, when issues arose some decades ago about the need to grow more gum trees and broaden the scope of plantations, it was possible to shed light on the issues by examining the oldest and best kept track record—the case of India. The inclusion of eucalypts in India's national and state schemes for rural development provoked great controversy. Many of the current concerns about the genus, which are being discussed in Latin America, have already occurred in India, where eucalypts are growing on more than 4.8 million ha (1.9 million acres).

Plantings in Other Nations

By 1900 approximately forty nations had accepted eucalypts as wood-producing exotics. That number has more than doubled over the past fifty years as people accepted the tree's multifaceted uses. Today, forest plantations intended for pulp dominate markets for eucalypts outside Australia.

Spain is an important grower of eucalypts. These plantations for pulpwood are growing near Vares, A Coruna, in Galicia. (Photo: Andrew Miles)

A plantation near Burela, Lugo, in the province of Galicia. (Photo: Andrew Miles)

Eucalypts surround the church of San Miguel de Lillo, Asturias. (Photo: Andrew Miles)

This eucalypt plantation grows in sand dunes beside Morocco's Merja Zerga, south of Kenitra on the Atlantic coast.

Eucalypts provide shade and poles for local communities around the Merja Zerga Nature Reserve, Morocco.

A hunting lodge stands amid fast-growing eucalypts in northern Morocco.

Roadside eucalypts on the road to Heptis, Tripoli, Libya. (Photo: R. K. Holz)

Eucalypts on the banks of the River Jordan, south of Lake Tiberias. (Photo: R. K. Holz)

Large, ornamental shade tree in Guanajuato, Mexico. (Photo: J. Sundberg)

Top left, *E. gunnii* (cider gum), flourishing above 50°N latitude in Beverley, East Yorkshire, England, in which a colony of rooks *(Corvus frugilegus)*, built nests.

Colonial Adoption

Eucalypts for India

> Eucalypts have come to stay in India as com-
> mercial tree crops. They have started playing
> their significant role because of its typical
> geo-climatic characteristics eminently suited
> to Indian socio-economic environment.
>
> —D. N. TEWARI, director-general,
> Indian Council of Forestry
> Research and Education, 1992

The presence of eucalyptus in India dates back to the rule of Tippu Sultan (1782–99), who ascended the throne of Mysore (now Karnataka), and who
reportedly planted seeds supplied by foreign traders. This potentate devoted
time to architecture and the design of gardens, and in about 1790 reportedly
introduced several eucalypt species into his palace garden on Nandi Hills,
some 65 km (40 mi) from Bangalore. His first plantings were for ornamental
purposes. At that time, a significant portion of the subcontinent's native timber resources remained intact. Tippu Sultan died in 1799 while fighting the
British, who rapidly expanded their control over the subcontinent. After that
time local evergreen woodland, or *sholas,* began to fall to satisfy ever-increasing demands for fuel and lumber.[1]

Concerns about the rate of deforestation during the early nineteenth
century led colonial authorities to experiment with the Australian blue gum,
notably in the Nilgiri Hills of Madras Presidency (now Tamil Nadu state) and
around the hill station of Ootacamund (where in the public gardens one tree
grew to be 30 m [98.4 ft] high in twelve years). In 1833 a British military officer
named Captain Dunn is believed to have added blue gums to the Nilgiris by

planting seed from Australia, with the intention of growing the trees as fuel-wood. A decade later another army officer, Captain Cotton of the Madras Engineers, had success with the same species in that hilly terrain. Trials proved promising and resulted in the establishment of firewood plantations.[2]

First Plantations

In the late 1850s and 1860s *E. globulus* woodlands sprouted vigorously in sites in southwest and south-central India. In reviewing the status of this newly acquired plant genus, forester J. E. O'Connor reported in 1872 that "for many years the Government of India has been importing large quantities of seeds of the trees of this family. . . . These seeds have been widely distributed with the object of acclimatizing such useful trees in parts of India best suited for their growth." Scientists found the seeds easy to transport, handle, and plant; rates of germination were excellent; and the saplings grew rapidly. Officials characterized the exotic trees as excellent for firewood and timber and possessed of health benefits ("miasma ceases wherever it flourishes, and fever flies before its face," was a cryptic way of describing its reputed malaria-destroying agency).[3]

As the East India Company acquired more territory, executives ordered more plantings—for example, in Lucknow, Saharanpur, Dehra Dun, and Lahore. Woodlots and small forests of eucalyptus, some of which dated back to the 1840s, climbed over the hillier regions of Mysore and Madras. In the Nilgiris blue gum grew four times faster than teak and produced comparable timber, noted the commissioner of Madras, and the species had also done well at Ranikhet in the Northwest Provinces. However, blue gum did not fare so well in the humid lowlands—around Calcutta, Assam, or Bengal, for example—or even in the uplands of Punjab and Coorg, where the commissioner and other interested officials had expected better results. Support for eucalypts came from highly placed officials such as Sir William Thomas Dennison (1804–71), governor-general of Australia, who included them in his promotion of agriculture, public health, irrigation, and public works after he took up the appointment of governor of Madras Presidency in January 1861. Others became involved as well: district conservators, sanitary commissioners, horticultural experts, forestry officers, and other officials, including Ferdinand von Mueller back in Australia, who supplied seeds of "tropically-inclined" trees to the Royal Botanic Gardens in Assam. Approximately 105 species of eucalypts received trials in the two decades after mid-century, and at least half a dozen of them proved useful.[4]

During this period forestry in India began to become "professionalized." Before the massive deforestation that occurred under British hegemony, forest dwellers considered woodlands as communal property and integrated management practices with religion and myth in regulating their activities. This tradition ended, however, as British officials established control over wooded areas and proceeded to undermine local lore about the cutting and collection of timber. A shift into European methods was geared toward increases in wood production, the regulation of tree felling, and the suppression of small-scale use by forest dwellers or "tribals."

In 1864 the British set up an Imperial Forest Department. This government entity dedicated itself to ending what officials insisted were wasteful and haphazard practices of wood cutting. A year later the Government Forests Act empowered authorities to declare any wooded land as a government forest and to punish those who breached its provisions. De facto control of tribal customs and usages was thus complete.[5]

Language in the Indian Forest Act of 1878 defined what constituted a forest and extended centralized control over forested areas (encompassing 23 percent of the surface area of India). Basically, the government reserved for itself the right to dispose of forests and wasteland in order to check what officials considered to be individual self-interest and shortsightedness on the part of villagers. The law classified the nation's forests into reserved forests, protected forests, and village forests, with provisions for each category. It has been argued that the comprehensive 1878 act made local people into sources of cheap labor by circumscribing what they could extract from wooded areas and by turning what had been communal lands into state ownership. One author notes that "the loss of community ownership had effectively broken the link between man and the forest. The harmony of village communities was rudely shaken with derecognition of communal rights and introduction of a money economy." Additional laws subordinated forest lands to cropped areas, making it a priority to expand crops, often at the expense of native forests.[6]

Government control over the access to and exploitation of forests continued to be elaborated and expanded into the twentieth century. In the 1920s and 1930s provincial authorities assumed some of the powers previously vested in the central government and its Forest Department. But officials continued to regard forests as basic sources of commercial wood and tax revenues, and to regulate local industries and those people whose livelihoods depended upon forest goods and services.[7]

Eucalypts featured in both national and state forestry schemes because of their extraordinary rates of growth, adaptability, and resilience, as well as the range of uses to which they could be put. Eucalyptus wood intended for railroad ties was, in general, a disappointment, as the cut trunks tended to split and shrink; nonetheless, the genus was well liked for providing poles, posts, lumber, and, of course, firewood. Trees took hold on a variety of sites, so that by the first and second decades of the 1900s a traveler would commonly encounter various eucalypts across the more elevated areas of southern India.

In Karnataka, for instance, lemon-scented gum *(E. citriodora)* grew quickly enough that people prized it as a fuel for the Mysore Iron and Steel Works at Bhadravathi in the Western Ghats. Forest red gum *(E. tereticornis),* known also as "Mysore hybrid," proved useful as firewood and timber around Mysore. This gum tree, also referred to as *E. hybrid,* is famous for growing on a range of soils—alluvia, *tarai,* laterites, sand dunes, red sands, loams, *murum,* and cotton soils—in more than a dozen states. Seeds from some of these gums, reportedly planted in the Nandi Hills in Tippu's era, did especially well as plantation trees. Of the 170 eucalyptus species tested in India, Mysore gum (with a Nandi provenance) has done best of all.[8]

Forest Policies after Independence

After India gained independence from Great Britain in 1947, eucalypts figured in a series of Five-Year Plans aimed at industrial growth and economic development. Forestry experts, who were part of a technocratic elite trained in colonial education systems, needed to grow wood as intensively and quickly as possible in order to meet increased demands for timber and firewood. According to these experts, who believed strongly in scientific principles bringing about European-style progress, the adaptable Australian plant genus seemed to literally spring up on barren as well as fertile sites. They counted on many eucalypts to provide usable wood within six or ten years. Gum trees did especially well in degraded areas, such as worked-out tea plantations or old cotton plantations. Eucalypts also fitted local needs by providing shade over roads, anchoring banks and levees along canals and rivers, and sheltering exposed and beaten-down ground in and around villages and on the outskirts of towns. It is no surprise that foresters favored them for economic and practical rea-

sons. Most scientists believed they upgraded the generally much slower growing native woodlands and had few qualms about replacing indigenous species
with these exotics.[9]

Scientific support for establishing plantations by setting out eucalypts
in large, densely packed geometric rows received a boost from India's Ninth
Silvicultural Conference in 1956. Participants recognized the trees' importance
worldwide and their excellent record of acclimatization and overall utility in
India. They urged colleagues in many states to begin "intensive cooperative
research" and to "streamline procedures for obtaining seed and data appropriate to each species." The conferees also instructed India's Forestry Research
Institute at Dehra Dun to publish details of the various methods and techniques for setting up and managing eucalyptus woodlands.[10]

At that conference, mention was made of the importance of a pioneering book about eucalyptus that had been published the year before by a French
forester named André Métro. A commentator believed that the publication
had "rendered a signal service in the dissemination of information regarding
the genus Eucalyptus and its various species." International interest and support drummed up and disseminated in similar silvicultural or forestry meetings further integrated this Australian plant into forestry planning and theory
in India.[11]

Study tours and international symposia sponsored by India's Forestry
Research Institute also spread the good name of eucalypts during the 1950s and
1960s. Enthusiasm over eucalypts and praise for their good qualities appeared
boundless. Experts encouraged an increase in both the number and size of
plantations, particularly after a government report in 1960 spelled out new production quotas for timber. What was needed was high production in the short-
est possible time. Foresters believed that eucalypts were perfect trees through
which they could realize such goals.[12]

The Tenth Silvicultural Conference in 1961 underscored this intent, urging that "all States would do well to use suitable species of Eucalyptus increasingly, particularly in their plantation and afforestation programs."[13] Three
trees were singled out. Mysore gum was appropriate for coastal sites and low-
lying places inland; flooded gum *(E. grandis)* would serve for "regions of
medium elevation . . . and good rainfall"; and blue gum was useful at "higher
elevations (above 2,000 m), with good rainfall (over 1,800 mm), but not liable
to snow-fall."[14]

Participants in the 1961 conference discussed ways to fight termites, drought, and early mortality among young stands of eucalypts. A few years later another group, assembled for the Eleventh Silvicultural Conference, pursued this direction of increasing wood output by pressing for additional research into eucalypt plantations. They looked forward to the establishment of the preferred species on a "large scale." All three eucalypts, the scientists declared, had adapted well to India and, most significantly, were able to satisfy an increasing market for paper and pulp. Conferees encouraged foresters to explore rates of tree growth under different conditions of soil and moisture.[15]

By the 1970s it was clear that regional support for these nonnative trees was enthusiastic and sustained. The First Forestry Conference at Dehra Dun in December 1973 featured nine papers on eucalypts. The published proceedings stated that "in view of the mammoth response from various states to the growing of eucalyptus and research in its various aspects," conference officials recommended that eucalypts should be "handled by separate research centers." More and more people, it seemed, were so enthusiastically committed to growing these exotic trees that regional government centers were needed to supply practical details about germination, growth, and transplantation. These centers should conduct trials for the most promising species.[16]

By 1980 tree specialists were busily holding trials of more than a hundred species and varieties in many localities across India. The Karnataka Forest Plantations Corporation raised *E. hybrid* on more than 20,000 ha (49,420 acres) in four forest divisions and planned to expand pulpwood by 1,500 ha (3,700 acres) every year. Foresters had also intensified research on the three well-established eucalypts, and they had added river red gum *(E. camaldulensis)*, lemon-scented gum *(E. citriodora)*, and swamp mahogany *(E. robusta)* to the list of important species. The Australian Development Assistance Bureau supported the introduction of high-quality seed from Australian provenances and promoted visits by Australian consultants to conduct inspections and to advise on how to make the transition from experimental trials to large-scale industrial plantations.[17]

By that time man-made forests blanketed more than 1.5 million ha (3.7 million acres) of India, of which at least a third consisted of eucalypts. About 441,000 ha (1 million acres) consisted of "block plantations," where eucalypts were grown at a density of 2,000 plants per hectare, mostly for pulp, after crops had been sown between rows for the first two or three years. Uttar Pradesh (82,000 ha [202,000 acres]), Karnataka (75,000 ha [105,325 acres]), Madhya Pradesh (51,000 ha [126,000 acres]), Tamil Nadu (40,000 ha [98,840 acres]),

Kerala (32,000 ha [79,000 acres]), and Gujarat (30,500 ha [75,365 acres]) headed a list of eighteen states in which people were growing eucalypts. This areal extent, however, did not include rows of trees beside roads, railways, or canals and lines of trees around fields and on village woodlots.[18]

Some regions experienced considerable expansion of plantations at that time. In Uttar Pradesh, where eucalypts were started in 1962, they had spread over 31,000 ha (76,600 acres) in 1969 and a decade later covered 82,000 ha (202,600 acres). They were the most important tree of nine broad-leaf species promoted by the state forestry department. Before eucalypts were to be planted, officials leased out the plantation land for field crops in order for farmers to work the soil and clear off native vegetation; then they turned the land over mainly to Mysore gum on a six- to ten-year cutting cycle. The aim was to help local farmers generate extra cash from eucalypt pulp. However, the scheme did not fare as well as expected. Tree survival peaked at merely 36 percent; crop yields decreased around field borders turned over to eucalypts, and growing trees shaded out crops. Similar experiments to grow eucalypts and field crops together met with the same problems in the 1980s.[19]

Today, forests in India occupy about 64 million ha (158 million acres), or approximately 20 percent of the land area. This expanse has declined from about 40 percent of land area in the last fifty years. People have tapped forests for many resources, including fuel, foods, fibers, and pasturage for livestock. Wood is the primary fuel for a rural population that exceeds 300 million. Recent estimates suggest that forests can yield 41 million m^3 (1.4 billion ft^3) of fuelwood annually, although current demand exceeds that figure sixfold. Hence it remains important to the Indian government that the area under tree cover be expanded, especially since domestic consumption of both fuel and paper has doubled in the last twenty years.

During the mid-1980s decisions made by the Gandhi administration to bring 5 million ha (12.4 million acres) of additional land under plantations for fuel or fodder resulted in a National Forest Policy program to extend the total forest area by at least 10 percent. A 1991 survey suggested that a modest annual increase was actually happening. Certainly, the rapid rate of vegetation loss has slowed, and plantations are credited with assisting in this slowdown.

One way of expanding the total forested area is to revegetate "wastelands" (defined as eroded and degraded surfaces), which a parliamentary committee estimated at 175 million ha (432 million acres; more than half of the nation) in 1985. Accordingly, eucalyptus has figured prominently in the national initiative both to plant more trees and to replant degraded surfaces. The Wasteland Development Program was devised to spearhead this task and thereby supply fuel and timber to rural people enmeshed in a web of poverty. Resistance to browsing and the capability of withstanding limited ground fires have given two species, *E. hybrid* and *E. camaldulensis,* a popular name in wasteland reclamation.[20]

Social forestry programs are another aspect of how India has dealt with its wood crisis. Established under the fifth Five-Year Plan (1976–81), the program was aimed at satisfying the needs of rural people for fuelwood, fodder, and lumber. With international support, planners implemented schemes that increased the amount of fuel, for example, on public land, village land, and wastelands. The theory was to involve local communities in the establishment of fast-growing plantations. The new trees, especially eucalypts, would also help relieve pressures on native woodlands.[21]

India's Eucalypts

Half a dozen species of eucalypts dominate afforestation schemes as well as federal and state plans to improve the lot of rural people in India. Mysore gum *(E. tereticornis,* also called eucalypt hybrid or *E. hybrid,* after specimens grown from possible crosses with *E. robusta* or *E. camaldulensis),* springing up from one small stand in the Nandi Hills, forms the basis for commercial eucalypt plantations that exceed 4.8 million ha (11.9 million acres; about a tenfold increase since the 1970s). At least 500,000 ha (1.2 million acres) of plantations—and some say half of all eucalypts in India—belong to this species. Mysore gum grows from sea level to about 1,000 m (3,300 ft) around the coasts and on semi-arid rolling plains where there is moderate rainfall. The species, however, will not tolerate snow or severe and prolonged frost.[22]

Over the past thirty years Mysore gum has become an important wood product in nine states, with Uttar Pradesh (82,000 ha [202,600 acres]), Karnataka (71,000 ha [175,400 acres]), and Madhya Pradesh (52,000 ha [128,500 acres]) outstripping competitors. It has also achieved the status of being considered an "Indian" tree because it was one of the first to be introduced, thriving best in the state after which it has been named. Mysore gum coppices well, regenerates naturally, and has received trials in seventeen states, more than any other member of the genus. Like conspecifics, it grows rapidly, adapts to a range of biophysical conditions, and has a wide variety of uses.[23]

Flooded gum *(E. grandis)* has grown well on at least 26,000 ha (64,250 acres) in Kerala and Tamil Nadu, and foresters have tried it in seven additional regions, including Assam, Orissa, and West Bengal. In general, this species prefers deep, well-drained, alluvial or volcanic soils up to about 1,000 m (3,300 ft) above sea level, with moderate precipitation. It continues to do well on barren and exposed sites in Kerala, where it was first introduced in 1948, and it looks promising on hilly lands in north India. Like Mysore gum, this species does not survive in heavy or swampy soils, but it has been used in social forestry programs and in plantations grown for pulp.

E. globulus was the earliest gum tree in India (see appendix 2, table 2) and has been the longest under plantation management, spanning about 11,000 ha (27,200 acres) in 1975. Seven states have been its locus in India, the most important being Tamil Nadu, including the Nilgiris and Palani Hills, with about 9,000 ha (22,200 acres). Over the past twenty years, the blue gum has spread into the uplands of south India, where it does best on well-drained soils in a cool tropical habitat up to about 2,500 m (8,200 ft) above sea level. Experts consider this species to be the fastest growing tree in India, and it is able to regrow by coppicing some four times in a ten- or fifteen-year cutting cycle before growers must replant. On good sites blue gum is also the most productive species, generating an average 30.3 m³ per hectare per year (433 ft³/acre) on the best spots. It is used in the manufacture of rayon and paper-grade pulp, as firewood, and as a source of cineole.[24]

River red gum *(E. camaldulensis)* thrives in semi-arid regions, and can withstand drought and mild frosts as well as poor soils.[25] In 1980 it was growing in fifteen states and territories. The red gum has grown well on reclaimed lands that have been mined for bauxite, as in Madhya Pradesh, for example. Its rapid growth and suitability for degraded conditions brought the tree to the attention of Prime Minister Rajiv Gandhi, who included it in his 1985 revegetation directive, aimed at putting 5 million ha (12.4 million acres) of land under fuelwood and fodder plantations every year. One of the main roles of river red gum has been to revegetate paths and tracks, and help stabilize sand dunes. Consultants have concluded that the most consistent high-yielding provenances have been from northern Australia, particularly from Petford, Queensland, and Katherine, Northern Territory.[26]

Swamp mahogany *(E. robusta)* and lemon-scented gum *(E. citriodora)* are the other species about which experts have expressed confidence. Both are fairly recent successes. The former has undergone successful trials in nine states, and the latter, which needs to be planted in a belt of rainfall between 800 and 1,500 mm (30 and 60 in) anywhere from sea level to 1,500 m (5,000 ft), is doing best in twelve states.

Environmental and Social Problems

The various wood-growing schemes launched in the early 1980s led some scientists, environmentalists, and interested persons to criticize the ecological and social impacts of India's forest policies. Opposition surfaced in 1983 in a series of articles published in *The Ecologist,* which challenged the paternalism and singlemindedness of foresters, most notably in respect to eucalyptus. Specific charges involved those scientists and officials who, under the aegis of the World Bank, had encouraged expansion of plantations not merely over wastelands, but into community woodlots and private lands.

In their first salvo "Eucalyptus—A Disastrous Tree for India," Shiva and Bandyopadhyay, two resource policy specialists, challenged the nation's social forestry program on three grounds. First, maximized growth of eucalypts for firewood was failing to meet other wood-related needs. Second, densely packed trees reduced overall biodiversity and contributed to environmental instability. Third, the switch into quick-growing eucalyptus was affecting village solidarity by encouraging a shift toward a cash economy. The net result, these opponents argued, was to channel hostility toward this genus and a sense of disappointment about the basic premise upon which social forestry was founded— namely, to improve the living conditions of rural people.[27]

In regard to wood, Shiva and Bandyopadhyay articulated what has since become an oft-repeated position: that although eucalypts do indeed satisfy measurable criteria for adding quantities of new wood within a short time, they fail to satisfy other practical needs. Rural people use woodlands for fuel, but also for building materials, utensils, herbs, edible fruits, and fibers supplied by a range of indigenous species. Reliance on eucalypts cut people off from this spectrum of benefits. Stockmen, for instance, noted that their animals refused to consume gum leaves. The exotic trees do provide fuel and small stakes for fences, but not edible fruits, fibers, or forage. People also grumbled that eucalypts are heavy users of surface and soil water. Local residents accused them of drying up the land, and of adding little in the way of fertility due to the pungency of their leaf litter, which reportedly inhibited the development of useful grasses and shrubs. Dislike of eucalypts grew, intensified, and spread.

Concerning biodiversity, Shiva and Bandyopadhyay and others noted that tree cover without a diversity of species to supply a variety of foods and fibers reduces the options and opportunities of local residents. This is why some farmers refer to the geometric blocks of monocrop trees as a "green desert." Linked to this view is the belief that, even crowded together, eucalypts do not prevent soil erosion. Interception of precipitation by their thin, pendulous leaves is merely a fraction of what native woody plants or acacia forests collect.

In respect to solidarity, Shiva and Bandyopadhyay and more recent critics accused officials of encouraging private landowners to expect sizable profits from coppicing eucalypts for pulp. Government projections exaggerated these potential cash returns, they insisted, and foresters and officials in turn encouraged individuals to thwart their roles and responsibilities toward the villages in which they lived.

Shiva and Bandyopadhyay concluded that rather than solving rural poverty, the expansion of the alien trees had compounded the problem. By promoting eucalypts, state forestry scientists were actually making those whom they were supposed to help more dependent upon state and national relief. The researchers appealed for a broadened perspective, insisting that "forest management in India has to stop providing one-dimensional solutions based on one-dimensional understanding of problems and evolve to a higher level of systems analysis and understanding."[28]

Events in Karnataka in 1983 illustrate the sense of frustration and alienation discussed by Shiva and Bandyopadhyay. People in several villages marched into plant nurseries, pulled out eucalypts, and in some cases replaced them with tamarind seeds. They were angered by the usurpation of village lands for eucalyptus plantations intended for the pulp industry. They made it clear to state foresters that they believed eucalypts deprived them of forest products, including firewood, which was traditionally taken as twigs and branches. Straight-trunked eucalypts had few limbs available to the villagers. With foreign support and encouragement, state officials were subsidizing the expansion of paper and rayon plantations. Outraged by new restrictions, people took the issue of eucalypt plantations on village lands to India's Supreme Court, claiming that fifty thousand inhabitants of seventy-six villages stood to lose from the social forestry schemes that were supposed to enhance their livelihoods.[29]

In Karnataka eucalypts had been planted from the 1950s under schemes for "fast growing species" and the "conservation of degraded areas." The U.S.-based Environmental Defense Fund criticized agencies such as the World Bank and the Overseas Development Agency for placing so much emphasis on eucalypts. Three years into the multimillion-dollar project for the state's social forestry project, the World Bank began to back away, urging a "considerable reduction" in the area given over to the alien trees. The bank's midterm review recognized that "too much emphasis was placed on physical planting targets," at the expense of meeting the forestry needs of rural people and of expecting the state of Karnataka Forestry Department to deliver such services to the poor. However, the chief forester at the World Bank defended eucalypts, stating that criticism was due "mainly to inadequate investigation of the scientific and socioeconomic evidence relating to the advantages and disadvantages of eucalyptus planting."[30]

Other reports disputed this position. One titled *Social Forestry in Karnataka,* authored by three Indian social scientists, noted that in Karnataka alone more than 100,000 ha (247,000 acres) of cropland had vanished under alien gum plantations between 1976 and 1981—that is, before the World Bank–sponsored scheme had gotten under way. Without reform the World Bank program (1983–88) would have added another 220,000 ha (544,000 acres) to this total and rewarded those larger landowners who could afford to convert a portion of their land to eucalypts for sale to the pulp industry.[31]

Wood-related schemes included in social forestry tended to rely on economic parameters—as was the case in Karnataka, for example—to the exclusion of ecological and social concerns. In testimony before a subcommittee of the U.S. Senate Committee on Foreign Relations in July 1990, Lori Udall, counsel for the Washington-based Environmental Defense Fund, noted that development schemes "have failed to adequately involve local communities and the rural poor and instead have catered to urban and commercial interests through the widespread promotion of fast growing tree species for pulp and paper and other commercial activities on private farmlands, wastelands, and community lands."[32]

N. C. Saxena paints a similar picture in researching the planting of eucalypts in six villages in Uttar Pradesh—a major beneficiary of social forestry initiatives—in north-central India during the 1980s. The scheme to plant eucalypts, he noted, had arisen after the National Commission on Agriculture recommended that government forests should be replanted with quick-growing species, and also that farm and village lands should be made more productive, again by planting with eucalypts. Consequently, in the first years farmers set 10.5 billion trees on private lands, mostly to generate marketable wood. The trees were cheap and required no irrigation. Most of them lined field boundaries or were planted as woodlots on and around tilled sites. Planters regarded them as a cash subsidy that would supplement returns from the arable land; harvesting of the young trees was expected to be flexible.[33]

When eucalypts were cut, they glutted the market and failed to meet expected cash returns. Rather than strengthening traditional agro-forestry practices, the rural tracts covered by blocks of eucalypts closed off options for browse, mast, and food, and those trees planted as fence lines or as smaller woodlots decreased crop yields by robbing them of moisture or casting too much shade.[34]

One problem found by Saxena was that farmers did not know how to include trees in their planting schemes, especially in central and eastern states given over to cultivation of household millets and rice. According to his analysis, most eucalypt projects did not increase family incomes. Only in western and southern regions, such as Gujarat, Karnataka, and Punjab, known for owner cultivation of cash crops on larger holdings, did investment in rotation tree crops pay off.

Social forestry as practiced in India between the mid-1970s and mid-1980s resulted in a massive tree-planting campaign, totaling hundreds of millions of individual trees. Subsidies for this campaign, which cost tens of millions of dollars supplied by international, national, and state agencies, helped decrease the rate of deforestation in the nation from about 1.5 million ha (3.7 million acres) per year to less than one-tenth of that rate. But short-term rotation plantations failed to realize the objectives of farm forestry. After cutting, eucalypts gained neither the markets nor prices expected by promoters; some smaller landowners were forced from their lands by encircling plantations, and abandoned cut-over plantation lands were exposed to degradation.[35]

Addressing the sociological framework for tree planting in forestry, Michael M. Cernea, a policy analyst for the World Bank, noted that between 1977 and 1986 about half of the World Bank's forestry lending went into twenty-seven projects involving community forestry, notably in India and Africa. Other international agencies poured resources into models for community woodlots, many of which involved eucalypts. The assumption was that local people would plant and conserve trees on communal lands. Expectations fell far short of being realized, however, because community forestry was assumed to be organized authoritatively, as it would be in China or Korea. Donors did not understand how social stratification within villages worked, how flustered and uncertain individuals became about benefits, and how they lacked knowledge of the skills and techniques needed for growing trees. There were distributional problems and a paucity of user-created management. In fact, state forestry personnel would frequently take control of operations and woodlots, selling off products outside the communities and thus engendering fear and mistrust.[36]

Agricultural economist Pranab K. Bhattacharya is one of those who support a more specific, piecemeal approach to eucalypt cultivation by planting trees in arid areas. He argues that they have succeeded in portions of Haryana, a scantily forested state next door to Uttar Pradesh. There, lands unfit for crops grew four species (*E. hybrid, globulus, camaldulensis,* and *tereticornis*). On the sand dune border with Rajasthan, he observed, the "tall sturdy eucalyptus trees were helping local cultivation by standing as wind breakers and consequently reducing wind velocity." He speculated that plantations in the Fatehabad division helped improve soil fertility. The unique aspects of rapid growth, tolerance of saline soils, and desert conditions make eucalypts an attractive proposition for the arid and semi-arid sections of Haryana where few plants survive.[37]

Vinayakrao Patil's account of the Nashik District Eucalyptus Growers Co-operative Society is similar to the story from Haryana. Founded in 1983 in Maharashtra state, the society succeeded in establishing eucalypts in an area subject to drought and risks associated with rainfed agriculture. About 15 million trees on 4,200 ha (10,400 acres) generated poles, firewood, and charcoal, for which Patil's organization found lucrative markets. Incentives through India's Wastelands Development Board, assistance from state foresters, and the willingness of farmers to share superior stock resulted in integration of crops with tree farming, Patil concluded. This policy of tree and crop diversification succeeded when conventional crop growing had so often failed.[38]

Haryana's tree-starved landscape, in which social forestry succeeded in vegetating 67,000 ha (166,000 acres), mostly with eucalypts, and their success on a smaller, drought-prone section of Maharashtra differ from the environment that Saxena encountered. Clearly, dry or droughty sites unable to support conventional crops may be put under gum species. In the first instance, officials regarded eucalyptus as a fuel (an alternative to kerosene), not as cash-producing pulp, as in neighboring Uttar Pradesh. There, Saxena argued persuasively, rather than encourage underemployment through a shift to these cash-crop trees, the government should promote food crops. Efforts to date, he believed, had rewarded larger landholders, who saved on wage labor by using eucalypts. The poorest rural families were left poorer, with neither rural jobs nor access to native woodland plants.[39]

Still, there are sites, such as ravines, sand dunes, desert fringes (as on the Rajasthan border with Haryana), and salty shorelines, in which eucalypt plantations, notably Mysore gum, may grow profitably. Little controversy exists about "greening" such arid environments. There is much more concern about planting eucalypts in so-called degraded forest lands, such as thorn scrub and grass savannas. Currently, almost 31 million ha (76.6 million acres) of the 75 million or so (185 million acres) officially recorded as "forest" have less than 40 percent tree cover. Argument centers on how best to revegetate this sparsely timbered and degraded surface in order to generate new wood and offset imports of both pulp and timber, which exceeded $420 million nationally in 1994.

The government has suggested turning many tracts over to commercial planting. In this scheme eucalypt plantations would be established on areas with less than 10 percent tree cover, or about 10 million ha (24.7 million acres). Only 2.5 million ha (6.2 million acres) of this impoverished land would be leased out for agro-forestry, and there the timber industries would work with the Forest Department (which retained ownership of the land) in order to plant native as well as commercially valuable species. This plan would not include forest lands in which local people have traditional rights.

However, nongovernmental organizations and others are not sanguine about this suggestion. They insist that the track record of the national government in leasing out rich forest lands to paper mills, which have worked them with eucalypts, has been neither an ecological nor an economic success story. "The proposal to co-opt industry in the task of rehabilitating India's forest resources needs to be carefully thought out," declared the business daily *The Economic Times* in 1995. Other ideas involving interspersing food and tree crops, increasing biodiversification, and replanting degraded lands with native rather than introduced species are part of the ongoing dialogue in response to the dilemma of meeting human needs from an increasingly impoverished biophysical base.

"The thoughtless conversion of natural forests into a monoculture of a handful of not very attractive species" is the basic reason for hostility toward eucalypts, concluded Karnataka forester Dilip Kumar in 1991 at an international gathering to analyze water and nutrient use, growth rates, and soil structure. Undoubtedly, he is correct. However, his appeal for a clearer understanding of the roles "played by trees in the economy" falls too short. It is not solely an economic issue, as G. N. N. Prasad and S. R. Ramaswamy of Bangalore insisted at the same meeting, which sought to address criticisms heaped on gum species, nor was it perhaps as "thoughtless" as the state forester suggested. In their judgment, it was the "deep-seated tendency of the Forest Department to look upon the [local] people as their adversaries." The two Bangalore-based nonforesters argued that really there was no need for additional research about "appropriate" trees or for further discussion about the biophysical constraints of eucalypt plantations. Rather, it was a matter of attitude — the unwillingness of experts to heed their own scientifically based admonitions going back twenty-five years. In their view, it came down to the patronizing tendency of authorities and experts to assume control of schemes that ultimately rewarded other people than those for whom they were devised. One official was quoted as admitting that most of the wood actually grown in social forestry schemes went to industry and urban centers, not to the rural poor. Equity over benefits continues to dog the eucalypt debate in India and elsewhere.[40]

Certainly, the debate about eucalyptus has brought these environmental and social issues to the fore, and seasoned campaigners are determined to fight against giving degraded lands — or any lands, for that matter — to industrial interests that will further degrade them. Misappropriation of wood in forestry schemes has hardened a constituency of nongovernmental organizations in India against eucalypts. Yet it seems reasonable to recommend that already degraded lands be planted in such fast-growing trees, while existing arable and wooded areas be left intact.[41]

Environmental Issues

Attention to the precise objective and
expected outcome of growing the trees (including social, hydrological and soil factors),
and careful choice of the trees to be grown
(indigenous or exotic, eucalypts or not, as
well as the particular species) is much more
appropriate than to use the trees themselves
as scapegoats. —A. G. BROWN,
CSIRO Division of
Forestry, Canberra, 1993

Over the past twenty years, institutional support for eucalypts has frayed
in India and in other nations whose residents had greeted them with initial
enthusiasm. Scientists and government officials are on the defensive, being
forced to explain and justify schemes that have contributed to the multiplication of large plantations and an expanding land area given over to these exotic
trees. The basic argument for adding eucalypts is that man-made forests help
satisfy national, regional, and local demands for an array of wood items. Local
and grass-roots responses to the plant genus challenge that position, attribut-
ing unexpected and unacceptable ecological problems to this massive expansion. Eucalypts produce new wood at high rates to be used as fuel and lumber,
but opponents argue that forest plantations close down options for other
resources at the local level, including food, medicines, and forage, which rural
folk would extract from native woodlands. On the regional level, as we have
noted in the case of India, many plantation schemes have also failed to fulfill
the promises that experts originally made. In addition, there is mounting evidence that some eucalypt plantations may damage the environment. Concerns

about such impacts have been raised in several countries, notably in Brazil, India, Thailand, and on the Iberian peninsula.

The Controversy

The scientific paternalism of eucalypt experts offends those who regard the plant genus as a source of burgeoning problems. Patronizing commentary and a sense of "we know best" or "who are you to tell us?" typify the gulf that has opened between scientists committed to afforestation with eucalypts and some of the populace who have had to live with the nonnative trees. Some foresters regard questions about the biophysical impacts of growing eucalyptus, and the potential repercussions on other natural resources, as emotional and ill-tempered. Criticism of plantations under eucalypts is inextricably linked, in their view, to politics, not science, and to the aims of militants who are bent on undermining their authority. In this scenario, the harshest criticism comes from an environmental lobby that attacks their professional reputations while ignoring important breakthroughs and benefits from modern agro-forestry.

Referring to India, S. A. Shah, secretary of the International Tree Crops Institute, states that "what we need . . . is trees, trees and more trees." Like a starving person who requires nourishment, India is cultivating trees to satisfy its ever-increasing appetite for wood, so that eucalyptus "is a better tree than no tree." Opposition to eucalypts in this sense is akin to the attitudes of the Luddites of the Industrial Revolution, who smashed the machines that subsequently freed so many from drudgery and dependence.[1]

Two separate but related issues require discussion. One deals with the environmental concerns stemming from the establishment of plantations—that is, from the planting of demarcated, uniformly dense stands of even-aged, monotypic tree species. What special effects do plantations in general, and eucalyptus plantations in particular, have on the biological diversity and character of the sites in which they flourish? The second issue explores the specific effects of the genus *Eucalyptus* on the biophysical environments in which people have established it. Opponents blame eucalypts for a range of problems associated with depleted soil moisture and fertility, competition with other plants, and even climatic change.

Plantations

Many tree scientists have a high regard for plantations. Forest plantations are set out as uniform blocks of evenly spaced trees usually intended for
growing a crop of wood. Foresters favor them for several reasons. Speaking
about tropical areas, English-based forester Julian Evans expresses support this
way: "Put simply, wood can be produced quickly, often on land hitherto unused. Firewood and numerous other domestic products can be grown efficiently. And, by planting trees, shelter and shade is provided and ground cover
is re-established which will halt and eventually reverse the degrade *[sic]* of
much of the tropical forest environment."[2]

Afforestation means to plant trees on surfaces denuded of them for a long
period. *Reforestation* means to replace trees on lands from which they have
been removed relatively recently. Whether they are intended for afforestation
or reforestation, tree plantations share several characteristics—and some disadvantages.

Plantations are intended to be virtual monocultures for which there are
predetermined uses. Like any field planted with a food crop, plantations consist of trees that exist to be harvested on a regular cycle in order to meet specific
requirements for lumber, fuel, or fiber. The purpose is to produce as much
wood of marketable quality as possible on the smallest unit of land in the short-
est possible time. Current estimates are that it would take about 150 million ha
(373 million acres) of fast-growing plantations (equal to the combined area of
Germany, France, Spain, and Portugal) to meet the global demand for wood.
In this view, tripling the current area of man-made forests under teak, pines,
acacia, eucalypts, and similar woody plants would go a long way toward accom-
plishing this objective.[3]

As monocultures, grids of similarly aged trees are not reservoirs of biodiversity, nor are they intended to be. Plantations are normally less diverse
than the plant communities they replace. The vegetation profile and structure
of plantations differs from that of indigenous tropical forests, which often
consist of variously aged individuals of many species, which do not usually fit
easily into harvesting schedules.

Plantations in the tropical areas of the globe (between 27°N and S latitude) are rapidly multiplying in number and areal extent. Evans cites FAO statistics that in 1980 11.5 million plantations covered 21 million ha (52 million

acres) of the tropics and that over the next ten years they doubled in extent (to almost twice the size of the United Kingdom). The tropics encompass about 40 percent of the earth's land surface and are the home for 50 percent of world population. A major portion of that population lives in poverty with scarce access to basic resources.

Thus, one important reason for establishing forest plantations is to improve the lot of the rural poor. Plantations grow wood products for domestic and international needs. At one economic level, plantation-derived industries offer employment; at another, they generate international exchange from the sale of pulps, hardboard, and lumber. Plantations also open up regional economies through advances in transportation, communication, and industrial diversification. They are a source of prestige, in the sense of being modern, productive, and efficient, for the nations that support them.[4]

Many schemes that have bolstered increases in the number and size of plantations have been aimed at counteracting rampant deforestation. Currently, as Evans explains, there is no dearth of wood in the world; in fact, there is enough to satisfy consumption for another seventy-five years without any concerted effort to replant. But spectacular and newsworthy accounts of indigenous forest loss, especially in tropical regions, have spurred national governments and international agencies into planning for more and more man-made forests as a way to decrease the felling of native trees.

Evans draws attention to the once-wealthy French colony of Haiti, one of the world's most destitute nations. The tree canopy on that island has shrunk to barely 4 percent of its aboriginal extent. Tree cover in Ethiopia has fallen to a similar percentage. Studies of India reveal that forest cover is barely 20 percent of what it was one hundred fifty years or so ago. Thailand has suffered the loss of half of its natural forests in barely twenty years, beginning in 1965. The list goes on, and the pace of tree cutting quickens. Efforts to select, plant, and manage forest plantations in these and additional nations slows the pace of woodland loss, which annually totals 7.4 million ha (18.3 million acres) in Latin America, 4.1 million ha (10.1 million acres) in Africa, and 1.9 million ha (4.7 million acres) in Asia.[5]

Another reason for the switch to plantations is that much more is known about how to grow a selected number of trees that will put on wood quickly, easily, and in large quantities, than about how to manage a natural forest on a sustainable basis. Eucalypts are among the earth's fastest producers of wood, and research has demonstrated that their yields can be extremely high. Flooded gum at Aracruz, for example, now achieves annual wood increments of 40–55 m³ per ha (572–786 ft³/acre)—more than double that of pines and araucaria. Growth rates among tropical eucalypt hybrids can exceed even the higher figure on a seven- to twenty-year rotation—approximately double the figure for tropical pines on a longer cycle and twenty times greater than for managed forests in the United States.[6]

There is ample "unused" land in the tropics on which to grow large quantities of cellulose. Brazil's interior Cerrado consists of 180 million ha (445 million acres) of natural, cleared, and grazed scrubland that foresters regard as suitable habitat for plantation forests such as eucalyptus. In such "empty" regions, land prices are low and the capital necessary for the investment is relatively small.

The Cerrado is one area among many that has been historically exploited and which scientists argue can be rehabilitated through both agriculture and plantations. In this scenario, eucalypts could assist in afforestation, returning long-degraded regions to better land health. At present there are at least 2.5 million ha (6.2 million acres) of eucalypt plantations in the Cerrado. With their emphasis on fast growth, high productivity, and felling trees before they attain maturity, plantation foresters recognize that plantations have a peculiar impact on the soils and moisture regimes in which they are planted. Frequently foresters resort to fertilizers to enhance growth and to pesticides for controlling ants and termites—resulting in pollution and impacts on indigenous flora and fauna.[7]

Diseases and Pests

The susceptibility of plantations to disease and insects concerns foresters and environmentalists alike. Any crop in large stands may become infested by insects, fungi, and viruses, which can swell to pandemic proportions. Once infestation occurs, the ample food source tapped by pests enables their numbers to grow so massively as to require pesticides, tree removal, or tree thinning—all of which are usually costly in both economic and environmental terms. In addition, once a disease gets a foothold, it may prove impossible to eradicate and can spread to adjacent tree or farm crops, and prove impossible to eradicate. The introduction and spread of pathogens is made easier among similarly aged trees, which are often crowded together on cut or cleared sites. Thinning out weak or defective trees and pruning dead or dying limbs helps reduce losses from pests, but these strategies also tend to lower expected yields. Natural pests may catch up with plantations in new locations, even though there is no evidence of them in the first trials. As a result productivity founders and public opposition builds.[8]

The issue of disease is particularly interesting for eucalypts. Insect infestations of eucalypt plantations have been comparatively uncommon beyond Australia. This is because most of the trade has been in seeds, originally from Australia and later from other established sources, rather than in seedlings or young plants. Seeds, which are tiny and easy to ship in bulk, have enabled commercial operations to minimize the risks of introducing pests. There are exceptions, however. The eucalyptus snout beetle *(Gonipterus scutellatus)*, widely distributed in Australia, invaded gum stands in South Africa in the 1920s. It fed on leafy shoots of the manna gum, for example, until introduction of parasites of the beetle's eggs curtailed damage, except in areas too cold for the parasites, such as Lesotho. The bug has also infested plantations in Kenya. During World War II, the eucalyptus borer *(Phoracantha semipunctata)* somehow arrived in Cyprus, Israel, and probably other nations in the Middle East. Canker damage to forests of flooded gum has also been costly in Brazil.[9]

One way of fighting pests is to pay close attention to provenance—that is, the source regions from which seeds originate. Healthy trees are most resistant to infestations, and some populations may have growth patterns, or may have evolved defenses, that ward off attacks. Guava rust *(Puccinia psidii)* lowers wood yields among several eucalypts in three states in Brazil, but trials with *E. grandis* seed directly from Australia, rather than from South Africa—the source region for many plantations—have had success in preventing losses. Similarly, the disease called Mal de Rio Doce, which has ravaged 30,000 ha (74,000 acres) of eucalypts in Minas Gerais, Brazil, and remains a serious threat nationwide, became much less damaging after eucalypt clones had been generated by vegetative reproduction from a single resistant tree grown from a seed lot sent from Australia. These clones have given such promising results that the species is recommended, with two other resistant species, for additional plantations.[10]

There is no guarantee, however, that pests that limit the development of gum trees in Australia will not eventually find and infest plantations overseas. History shows that sooner or later pests "discover" their favorite hosts even in habitats far outside the aboriginal range of a plant or animal. Other pest organisms may also adjust to the new trees. For example, as expected, termite damage is common in several regions in both Africa and India. Despite this, compared with other plantation species, eucalypts have performed well in resisting pests and diseases.

Wildlife

In general, plantations are inferior habitats for native animals as compared with native forests. Plantations are crops, and like other crops they are not intended to support wildlife diversity. One study in Sri Lanka, for example, noted that the number of bird species increased in tree-filled sites—from the lowest number in pine and eucalyptus monocultures to a higher number in home or dooryard gardens and to the highest number of all in a natural forest. The number of species on plots around houses compared favorably with those in the natural forest when fruiting plants were included.[11]

Biodiversity may be enhanced in plantations by adopting strategies aimed at conserving microhabitats in and around tree-dominated sites. Corridors of indigenous shrubs and trees left along rivers and streams, for instance, benefit native animals. These pathways may also function as the means of preventing soil erosion during times of flooding. The establishment or retention of herbaceous and shrub-filled openings, or of native plants interspersed throughout a forest plantation, also enhances food and cover for wildlife. Patches of native forest within or along the edges of large plantations also offer food and cover. In Aracruz, foresters stipulate that native vegetation must be included with nonnative eucalypts. Apart from the area retained or reforested in native plants, none of these management strategies will necessarily impinge upon the economic productivity of man-made plantations.

In addition, raising various age-classes of trees within the plantation and leaving some specimens to mature may be helpful to indigenous flora and fauna. Timing the cutting schedules to allow for periods of fruiting and flowering is another management option that may assist native mammals and birds.

Plantations established on denuded savannas, grasslands, or worn-down croplands will normally be expected to increase overall faunal diversity, although wood-dwelling vertebrates may replace grass-adapted ones. It was discovered in India, for example, that native deer, ground-dwelling birds such as bustards, and other rare animals returned and built up their numbers when plantations grew up on areas that were previously open and exposed. It is important, however, to leave portions of a grassland or parkland intact, if one intends to conserve indigenous species, such as the bustard, that are not adapted to dense tree cover. Emus occupy pine plantations in southern Australia and forage in fire-breaks. In this way, man-made forests may become refuges for wildlife, even in densely populated areas. There, native animals may seek out plantations as denning sites because people generally avoid the compact, gloomy woodlands in their daily activities. Freedom from disturbance and hunting thus enhances faunal diversity.[12]

In general, eucalypt plantations do not provide useful food resources for animals. Their small hard seeds are unattractive to birds. Deer and domestic stock find their foliage inedible, and the paucity of ground cover under the stands of many plantation species reduces the amount, type, and quality of forage. Improvements in plantation design, based on intercropping and the creation of vegetation edges, ponds, openings, and clearings, can compensate for the homogeneous character of plantations themselves.

In Australia, several well-known animal species thrive in eucalypts, including koalas, which feed on the foliage of about twelve species. Essential oils in gum leaves help control ectoparasites on koalas, making the cuddly "bears" smell "just like cough drops." Foods in the form of nectar, insects, and seeds from eucalypt stands, and rest sites supplied by cavities, branch forks, bark shoots, dead snags, and clumps of leaves in large, mature trees, are the mainstays for specialized Australian birds, including parrots, honeyeaters, and pardalotes, and predatory falcons, currawongs, and owls.[13] Outside Australasia, however, many animals do not adapt well to eucalypts. Reginald E. Moreau, a pioneer researcher in African ornithology, believed that eucalypt plantations were largely ignored by African birds. Others maintain that his statement was too sweeping. Although pure stands "afford a remarkably sterile environment," according to bird expert K. D. Smith, some African birds do nest in and among the trees. Local circumstances and the condition and age of plantations affect the presence and numbers of avians.[14]

Smith noted that on the largely treeless Highveld of Transvaal in South Africa, eucalypts provided substantial habitat for birds. Plantations supported about twenty species, as compared with about fifty species in the bordering bush; although the number of species was lower, several had extended their ranges due to the existence of eucalypts. In South Africa, Smith discovered that neglected and overgrown plantations held more species of birds than well-tended ones. Only one, the black goshawk, preferred the heart of large, mature, and well-managed woodlands, whereas many other species foraged along timber edges. Smith reviewed eucalypts throughout Africa and discovered that thirty-three bird species were known to nest in eucalypts in North Africa, twenty-two species nested in equatorial areas, and seventy species were noted in the south—that is, from Zimbabwe through South Africa. Most were waterbirds, such as colonial nesting herons or storks, which built their nests in the trees. Some were birds of prey, which used the gum trees as perches or as lookouts for potential prey. Hawks and falcons were attracted by the height of eucalypts, because in the tall trees their nests would be safe from ground-living predators. In all, Smith recorded a total of ninety-eight bird species breeding in eucalyptus.[15]

Another South African study identified similar niches filled by specialized birds in Australia that were not filled by native birds in South Africa. However, its author noted that some species visited plantations or used them during times of winter drought, and remarked that plantations of these exotic trees "were not as sterile and unsuitable for bird life as they were often accused of being."[16]

This opinion is in line with recent research conducted along the banks of the Rio Plata in Uruguay, where Jane Lyons studied two urban parks in and near Montevideo that were richly planted with eucalypts. She noted that 52 of 131 species (40 percent) observed in the preserves used eucalypts. Several insectivorous birds perched readily on branches while hunting insects; woodpeckers moved about the trunks, and hummingbirds and tanagers were attracted to flowers as sources of nectar. Nineteen species nested in eucalypt groves. Lyons concluded that eucalypts seemed to have none of the serious limitations typified by densely packed plantations. Some trees were old and very tall, others had a well-developed understory of native vegetation, and widely spaced plantings formed a mosaic of nonnative and native plants favored by many birds.[17]

The Montevideo study suggests that eucalypts may be attractive to birds when mixed with other plants, providing a profile in the vegetation that attracts a large number of species. In Australia, structurally complex eucalypt forest and woodlands support more birds than simple ones because the existence of vegetation layers at different heights provides both food and shelter. Some groups of birds glean the topmost leaves and branches, while others forage on the bark and large limbs lower down. Some species visit shrubs under the canopy for sources of nectar, and the remainder search the ground for seeds and insects. Parrots and other birds readily utilize cavities in mature and old trees as nest sites.

Compared with indigenous forests, plantations of gum species hold little attraction for birds because there is no layering of vegetation within the plantation. The limited variety of trees, their immaturity, and a poorly developed understory restricts the number of species. Food proves scarce, except when trees are in flower. At that time, insectivorous species prey upon bees and other honey-gleaning insects or feast upon the nectar themselves. Birds will also seek out gum trees as roosts or select them as perches in landscapes mostly denuded of taller woody plants.[18]

Studies from the 1970s show that outside Australia eucalypt plantations have a limited appeal, not merely for avians, but also for other wildlife. Research in Brazil discovered that small mammals were more abundant in a thirty-one-year-old plantation of native *Araucaria angustifolia* than in a ten-year-old plantation of Sydney blue gum *(E. saligna)*. Two evergreen tropical rainforests, recovering from exploitation, also supported higher numbers of small mammals than gum plantations. In a study of spider webs in Malawi, one researcher concluded that web density was far higher in *Brachystegia* woodland than in adjacent eucalypts. He believed there was insufficient food among the eucalypts for many spiders to live on and surmised that "[spiders] would disappear in large pure eucalyptus plantations."[19]

Although plantations in general have a less diverse array of wildlife residents, those consisting of eucalypts are even poorer because they provide less food and cover. Compensatory measures can be taken, however, to enhance their value to wildlife. Stands of native tree species may be left intact along waterways, or interspersed among plantations. In treeless areas, such as the Highveld in South Africa and the Andes of South America, the planting of eucalypts may actually benefit animal populations. Insects and birds are attracted to new nest sites and sources of nectar, and mammals may use fallen leaves and bark as cover and nest materials.

Biophysical Issues

There is controversy about whether eucalypts degrade the environments in which they are planted. Anecdotal evidence details how gum trees dry out soils, deplete groundwater, and promote changes in regional climate. Local people record how groves poison adjacent plants, promoting soil erosion due to the bare ground that is exposed under their canopies. Further criticism focuses on the ability of eucalypts to exploit nutrients at higher rates than other trees. Over the past twenty years local people in tropical and subtropical situations have accused the plant genus of being hard on the landscape and of radically altering the biophysical components of Mediterranean-type and subtropical environments in both the Old World and the New World in ways that close down options and opportunities for people in rural communities.[20]

Foresters and arboriculturists reject such accusations as unscientific and based on hearsay. Experiments demonstrate that eucalypts use about the same amount of water and nutrients per unit biomass increase as any other equally fast-growing plant cover. A 1984 study by the Forest Research Laboratory at Kanpur discovered that water consumption per gram of eucalyptus wood produced was less than for four indigenous woody species. "It is evident the Eucalyptus is much more efficient than several indigenous species in respect of water utilisation," noted the New Delhi–based Birla Institute of Research. "[If] more wood production is the objective, the cost has to be paid in terms of more water."[21]

A 1985 FAO report that reviewed data about the ecological effects of eucalypts concluded that site and locality, plus cultural traditions and practices, may play important roles in the impact that eucalypts make on the landscape. The authors of the report also noted that basic research about environmental impacts of the genus comes from a relatively circumscribed area, mostly from India and Mediterranean Europe.[22]

Water Use

The genus *Eucalyptus* has been the focus of serious concern in regard to its reported profligate use of water, which critics argue lowers the water table, robs soils of moisture, and limits the type and amount of adjacent or understory vegetation, including farm crops. Some scientists support this accusation, others do not.

In their review of environmental effects of the genus, M. E. D. Poore and C. Fries concluded that the "strong surface roots of some eucalypts mean that they compete vigorously with ground vegetation and with neighboring crops in situations where water is in short supply." Data from the humid tropics suggest that young, vigorous tree stands "consume more water, and regulate flow less well, than natural forests."[23]

Plant authorities B. J. Zobel, G. van Wyk, and P. Stahl agree, noting that "when the soil is moist, trees of this genera do, in fact, use more moisture than most other trees."[24] Forester Julian Evans also contends that eucalypts are not "drought evaders," reporting that stomatal closure often occurs late—that is, usually after moisture stress has grown severe. In general, transpiration rates may be high and directly proportional to photosynthesis rate and to growth rate. The trees have extensive root systems, and depending on the species, the root network may be both wide and deep enough to tap water supplies inaccessible to other plants. Thus, in times of drought, some eucalypts can do better than other vegetation.

Without competition, eucalypts tend to use soil moisture at high rates. According to Australian-based forester Ross G. Florence, it follows that high wood production is achieved through increases in both water and nutrient uptake. Although he did not conclude that the genus uses water inefficiently, Florence noted that some trees—for example, *E. regnans*—have accelerated rates of transpiration and apparently little capacity to regulate water loss. Very rapid growth demands lots of water and nutrients.[25]

Hydrologist I. R. Calder addressed the controversy in India about water use by eucalyptus at an international symposium in Bangalore in February 1991. He noted that in terms of interception—that is, the amount of precipitation captured by the trees and then lost through evaporation—eucalypts, like other woody plants, gather more moisture than shorter vegetation. In terms of transpiration, eucalypts are like other trees, except in some situations in which a species does not regulate stomatal closure, especially when soil moisture is abundant and atmospheric demand high. When stomata are open, there is carbon dioxide and oxygen exchange (photosynthesis) and loss of water. Then, eucalypts may become "marsh reclaimers" or "water pumps." Calder concluded that if eucalypts are planted in croplands, then moisture losses will increase due to rainfall interception and deeper roots that exploit subsurface moisture. But if they are planted in previously forested areas, there is little difference, except for those species given to high rates of transpiration. This position reflects studies in Israel, corroborated by the work of India's Forest Research Institute.[26]

On and above the ground, eucalypts intercept anywhere from 10 percent to 25 percent of the rainwater that falls through their leaves. As with any forest, this means that less water reaches the soil, and depending on the degree of slope, type of soil, and type of vegetation, runoff ranges from nonexistent to large amounts that require devices to control erosion. In general, experts recommend that where rainfall is less than 400 mm (15 in), food crops should not be mixed with eucalypts; in rainfall regimes between 400 mm and 1,200 mm (15 and 50 in), crops are sustainable provided tree density matches the desired balance of water for other uses; in regimes above 1,200 mm (50 in), no special precautions are needed.

Erosion

Although native eucalypt forests in Australia are prized for erosion control, eucalypt plantations are not especially effective agents for this purpose. In fact, critics believe eucalypts promote rather than inhibit soil erosion through the failure of narrow leaves to intercept a large percentage of precipitation. Gullying may occur in plantations on steeper slopes. Saplings may be crowded out by grass or shrubs, so that it may be necessary to cut back or remove weeds in order to facilitate growth. In such situations, exemplified by studies in Portugal, cleared patches around the young trees tend to wash out and scour. As specimens mature, the leaf litter they deposit is not dense enough to prevent runoff. Pastoralists and others accuse this litter from Australian trees of "poisoning" other plants beneath and around plantation stands. Bare ground under the trees not only may limit the range of fodder and edible fruits, but also may fail to contain or control water. Terracing, thinning out trees, and leaving ground clutter after felling are ways to offset unsuitability for erosion control.[27]

The issues of alleopathy—producing chemicals that stymie the presence of other plants—and water repellency also arise in arguments about woodlots and plantations. The question of whether eucalyptus trees poison and retard surrounding vegetation is the topic of energetic debate. Much of the experimental work that has been done involves laboratory studies that do not replicate conditions in the field. Water repellency is caused by a layer of microbes and fungi on the surface of the soil that impedes percolation. Although water repellency, like allelopathy, is not confined to genus *Eucalyptus*, this characteristic may inhibit the growth of a strong understory, so that when the trees are cut the exposed soil dries out and hardens, making it subject to gullying.[28]

The techniques employed in felling, coppicing, or clearing the site often dramatically accelerate soil compaction and loss around tree farms or plantations. This is obviously true for hilly areas. Unless preventive bunds or bolsters are set out, both surface flows and erosion will scour the slopes from which eucalypts have been extracted.

Nutrient Uptake

Another frequently raised issue is whether gum trees exploit and excessively deplete nutrients in the soils. It appears paradoxical that foresters select eucalypts because they flourish on poor soils and barren sites, and that the trees are then blamed for causing these substrates to become even more impoverished. Nutrient uptake and storage through the deep and widespread root system, and the efficient internal cycling of chemicals within the trees, add to eucalypts' reputation for flourishing on impoverished soils.

Selective pressures associated with the evolution of the plants in the mid-Tertiary in Australia may explain this capability to grow on poor and impoverished substrates. However, foresters also point out that the fastest growers are invariably on relatively fertile sites. Trees will exploit a nutrient pool when one is available or supplied for them.

The type and amount of nutrient uptake or removal also depend on the methods of harvesting the trees, the type and character of the wood itself, and the age at which coppicing or clear-cutting occurs. The younger the plantation, the larger the loss of nutrients after cutting. Because heartwood in many species forms at about seven years of age, that is when nutrient redistribution begins to take place throughout the stand on a more efficient basis. If cutting for pulp happens during those initial years of very speedy growth, nutrients are lost from the ecosystem. Indirect losses may also occur due to soil displacement during felling, burning of the litter or slash, and leaching—activities associated with exposed sites subjected to washout. Expert opinion suggests that eucalypts do not necessarily impoverish substrates any faster than other rapidly growing species that produce wood at comparable rates. Longer rotations, attention to site management, retention of bark and branches on site at harvest, planting of legumes, and overall tree husbandry can diminish or mitigate the losses of soil chemicals.[29]

Historically, foresters introduced various eucalypt species with little regard or appreciation for local conditions or for what could be expected realistically from the trees. Their objective was to help seedlings to survive and grow as much wood as rapidly as possible. Today, there is a far better understanding of the available genetic material, which earlier programs ignored. Frequently, schemes for the introduction of eucalypts failed to study what species would do best and in what numbers, and to test the breadth of their genetic base. Thus, introductions frequently used seeds from only one source (maybe a tarnished one) and grew different eucalypts together, thereby promoting hybridization and perhaps an ultimate loss of vigor. Problems occur when a species fits neither the site that people have selected for it nor the needs of those for whom it is being grown.[30]

A Global Solution

International Agencies Promote Eucalypts

As a tree for industrial plantations eucalypts
can be conspicuously successful, if the
species and provenance are well matched to
site and such plantations are an appropriate
form of land use in the locality.

—D. POORE, P. G. ADLARD,
AND M. ARNOLD, Commonwealth
Forestry Association, Oxford, 1988

In the 1950s international agencies began to support and actively promote the cultivation of eucalypts in areas in which they had never been grown
before, as well as in places in which they were already established. The new
regions selected for the trees were mainly in tropical and subtropical nations
in South America, Africa, and Asia. Agencies established trials and research
programs that poorer nations could not afford to undertake or lacked the facilities or expertise to support. The impetus to open up new areas to these exotic
trees and to accelerate the spread of species in other areas in which they were
well established was linked to international aid programs. These programs
were geared toward the production of wood (mostly for fuel), the restoration
of degraded environments, and the planting of trees on arid lands. In addition
to the essential support of international agencies, this rapid and unprecedented
expansion of range was also made possible through the work of individual
researchers in several countries.

Most of the individuals who pioneered the study of eucalypts worked
for federal or state agencies, including universities. This was true in both Australia and the United States. For example, the U.S. Department of Agriculture

and forestry organizations in California expended funds, time, and energy on the Australian trees. Officials advised the public about where and when to plant, how to grow the trees, and what to expect from them in the way of commodities. With ties to enthusiasts such as Ellwood Cooper and Abbot Kinney, the forestry establishment spurred both expansion and acceptance of the trees. As time passed, however, these officials began to reassess the plant genus and the plans and hopes they had expressed for it, modifying their earlier, unconditional support.

In California, as we have seen, initial plantings did well but were not an unqualified success in either economic or aesthetic terms. Carpenters and construction engineers grumbled that eucalyptus wood was too hard or too friable, or that it warped and split too much. Landscape gardeners disliked the untidy litter and limbs that built up under tree groves. Homeowners using eucalyptus for firewood said it burned too hot (some said too cold). Real estate salespeople and investors despaired about its profitability as a fuel or timber and began to seek other speculative ventures instead.

In Brazil—and to a lesser extent in India, where colonial officers decided that eucalyptus was useful in reforestation schemes—support was more steadfast. In Brazil, company executives took a shine to the Australian trees due to their adaptability and overall utility. Edmundo Navarro de Andrade's "monastic" dedication to the genus in the state of São Paolo appears never to have caused him doubts or scruples. His enthusiasm for trials and continued study remained constant.

After Navarro de Andrade's death, another generation of scientists followed the research path he had pioneered. The post–World War II experts were silviculturists, foresters, agronomists, horticulturists, consultants, and officials who often worked for or were funded by international agencies. The Food and Agriculture Organization of the United Nations (FAO), based in Rome, played an important role in eucalyptus research and promotion, and has continued to support the Australian trees. Founded in 1946, the FAO had among its objectives improvement in standards of living worldwide, especially of nutritional levels among rural people, and an increase in the overall productivity of agriculture. Forestry is an integral part of the FAO's technical and advisory programs on behalf of government and international funding organizations. Agency projects seek to establish new trees and crops, and include tree planting in schemes designed to meet the needs of agricultural infrastructure and development in various nations.[1]

Given the widespread desire to improve living standards and an unprecedented demand for wood by increasing numbers of humans in developing nations, it was logical that eucalypts should figure prominently in plans and publications calling for increases in wood production and the revegetation of deforested areas. Using data from the early 1950s, André Métro wrote a book called *Eucalypts for Planting* in which he explored planting, cultivation, and use. His 403-page book, published by the FAO in 1955, included short summaries of sixty-seven popular species, varieties, and hybrids. Métro, who was commissioner of water and forests at the École Nationale des Eaux et Forêts in Nancy, France, listed the tolerance of eucalypts to different regimes of temperature and moisture, and to various types of soil. His twelve-page bibliography showed the growth of literature about eucalypts. He made references to growth, disease, and coppicing, and mentioned oils, tannins, and honey, intending to promote the notion of eucalypts as a tree "crop" to administrators, agronomists, and others, especially in developing nations. The crop could supply traditional wood products plus resins, medicines, and food. Métro's volume became the benchmark against which later publications were measured.

In 1979, again with FAO support, a second eucalyptologist, Maxwell Ralph Jacobs (1905–79), penned a far more authoritative work. Jacobs was principal of the Australian Forestry School and director-general (1960–70) of Australia's Forestry and Timber Bureau in Canberra. In the late 1970s, when he was writing, eucalypts grew on 4 million ha (10 million acres) outside Australia, a fivefold expansion since Métro's time. The increase was the result of a significant growth of interest in the multipurpose trees, including use for fuel.[2]

Jacobs's 677-page book, also titled *Eucalypts for Planting* (published in the FAO Forestry Series), opened with a description of the "natural environment" of the genus in Australia, including range, growth characteristics, and the systematics of the entire genus. An important 300-page section of "species monographs" updated Métro's data, treating 117 species (almost twice as many as Métro had covered) that were then growing in ninety-two countries. In his foreword to the volume, E. Saouma, director-general of the FAO, declared that Jacobs's book, like Métro's, would assist foresters in choosing "the right tree to plant on the right site, using the right treatment."[3]

The "right" sites were on impoverished and intractable soils, particularly in the tropic and subtropics, which eucalypts could rehabilitate. Matching a species to a site was the theme of Jacobs's book. His twenty-nine-page bibliography reflected the increasing tempo of research and publication on eucalypts, and listed thirteen FAO publications on the subject since the first eucalyptus book had appeared twenty-five years earlier.[4]

Organizations linked with the FAO shared this enthusiasm for promoting eucalypts, especially those with which Max Jacobs had firm links in Australia. Almost 10 percent of Australia's twenty-five hundred tree species are commercially important at home or overseas. Acacia, casuarina, and that nation's only economically cultivated food crop, macadamia, all have their own specialized publications, which spell out criteria for site selection, planting, harvesting, processing, and use. In the trend to "domesticate" Australia's useful plants, foresters, agronomists, and others have singled out the genus *Eucalyptus* for prolonged, extensive, and cooperative research. Concerted efforts to improve the stock of selected species, especially among foreign nations, go back to discussions between the FAO and Australian government officials that led to Jacobs's call, at the Second World Eucalyptus Conference in 1961, to start a eucalypt "seed bank." Jacobs involved the Canberra-based Forestry and Timber Bureau, which he directed, in the Department of National Development in this move to centralize seed study, production, and disbursement. This function was transferred in 1975 to the Commonwealth Scientific and Industrial Research Organization (CSIRO), where it exists today as the Australian Tree Seed Centre (ATSC) within the Division of Forestry and Forest Products. ATSC supplies researchers and growers with quality seeds of eucalypts and other native woody plants, offers guidance to personnel and clients in more than one hundred countries, and continues to study the genetics of eucalypts and other species in order to upgrade varieties.[5]

ATSC experts established their credentials in the arena of eucalypt research in the early 1960s, when they supplied *E. camaldulensis* seed for performance trials under the aegis of the FAO's Mediterranean Forestry Research Committee. Currently, the program is influenced by priorities established by funding agencies, especially FAO-related needs and programs. In recent years ATSC staff have responded to some one thousand information requests per year, have offered training to foreign scientists, and have visited areas inside and outside Australia in which trials were being designed, initiated, or evaluated. In 1979, for example, ATSC eucalypt specialist Douglas Boland traveled to India, under the auspices of the Australian Development Bureau, to assist with supplying high-quality eucalyptus seeds, inspect ongoing trials, provide literature about appropriate species, and acquaint his Indian colleagues with

CSIRO services. Boland took a particular interest in *E. tereticornis* provenances from north Queensland, which experts believed were likely to exceed the quantity of wood products derived from *E. hybrid* (itself, in Boland's mind and in the estimation of expert colleagues, most probably *E. tereticornis*). The ATSC's contribution to provenance trials in India was part of a worldwide effort by the International Union of Forestry Research Organizations (IUFRO) to test eucalypts. About that time Australian authorities were processing annually more than three hundred orders involving three thousand seed lots for ninety nations. Since then the export market for seeds of Australian trees has risen to approximately $9 million annually (as of 1993) and shows no signs of declining. Such communications and visits, and the network of commercial and scientific contacts they generate and sustain, are a vital means for disseminating, formally and informally, information, data, and expertise about current and anticipated work on eucalypts.[6]

The Australian Center for International Agricultural Research (ACIAR), operating within the government's portfolio of foreign affairs and trade, assists developing nations with their agricultural problems. Through research partnerships with national institutions, it constructs the infrastructure for scientific research and applications. Established in 1982 to enable Australian scientists to assist foreign nations and resolve domestic agricultural issues, ACIAR commissions institutions, such as CSIRO and home universities, to embark upon partnership programs with developing countries to sustain and enhance agricultural production, train foreign scientists, and administer Australia's contribution to the international agricultural research centers. Recently, the agency has actively supported the evaluation of eucalypts for fuel, pole woods, timber, and other products in various nations, especially around the Pacific Rim. For example, a project aimed at improving and sustaining the productivity of eucalypts by increasing yields from *E. camaldulensis* in Southeast Asia has focused on Thailand, Laos, and Vietnam. The objectives are to breed more productive plant lineages, to increase total use of the trees (including extraction of leaf oils), and to explore the possibilities for growing gum trees together with nitrogen-fixing trees such as acacia in order to lessen the long-term drain on soil minerals. Working with their counterparts from those nations, Australian foresters concluded that provenances from Queensland, notably around Petford, performed generally better than seed stock from Western Australia or Northern Territory. The pilot study, completed in 1990, was declared generally promising, and the forestry departments in Thailand, Laos, and Vietnam all expressed support for continued trials and exchanges beyond the time constraints of ACIAR-sponsored trials.[7]

Currently, ACIAR-related projects involving CSIRO personnel are under way in China, where tens of thousands of hectares are being planted annually with acacias, casuarinas, and fast-growing gum trees. ACIAR has supported collaboration between Australia and China since 1985, and in 1992 it declared that the joint research "has had remarkable success." For their part, Australian-based foresters discovered that previously untested eucalypt species and provenances were doing better than earlier plantings in China; for their part, the Chinese foresters were learning how to turn trials and experiments into full-fledged plantations.

Although eucalypts in China can be traced back at least a century (see appendix 2, table 2), when *E. globulus* and a few other species were planted, a total of two hundred species, mainly belonging to the subgenus *Symphyomyrtus,* are known to have been tried at one time or another. In the 1980s, however, through active Australian support and partnership, at least 63 warm-area to tropical-type species from 258 provenances were tested, together with hybrid crosses (for example, between *E. camaldulensis* and *E. urophylla*) that have proven useful on less fertile soils in exposed areas, such as Guangxi Province. Currently, about 1.5 million ha (3.7 million acres) of gum plantations exist in China, and contacts between CSIRO scientists and their counterparts in the Chinese Academy of Forestry remain crucial to continued plantation growth. Collaboration includes studies of how introduced Australian root fungi may improve the establishment and growth of eucalypt plantations.[8]

Additional British Commonwealth and colonial associations have contributed to the growth in demand for eucalypts. Seven years after Métro's book appeared in 1955, R. J. Streets published *Exotic Forest Trees in the British Commonwealth,* an updated version of a similar work by R. S. Troup that had appeared in 1932. Streets's work was commissioned by the Sixth Forestry Conference of the British Commonwealth, held in Canada. At that international gathering, a standardized set of reports updating Troup's research was compiled. Working in the Imperial Forestry Institute in Oxford, England, Streets collated and published these reports, which were intended for "practicing foresters throughout the Commonwealth." Eucalypts figured prominently in the work.

Streets focused upon the economic value of eucalypts, including them

with pines, cypresses, teak, mahogany, poplars, and a number of Legumi-
nosae. He was impressed by how successful they had been in how many different regions. For instance, he noted that South Africa had a large area given
over to eucalypts, and that they were doing well in Zambia and Kenya. Eucalypts had become permanent members of New Zealand's flora. Understandably, the one area in which eucalypts had failed was in the British Isles, whose
25,000 ha (61,750 acres) of introduced plants were mostly conifers. Streets
devoted 110 pages to eucalypts, discussing twenty-four economically important species.[9]

World conferences devoted to eucalypts have served to sustain and expand interest in putting larger and larger amounts of land under these Australian hardwoods. The first such gathering, held in Rome in 1956, was the
product of a study tour of Australia, sponsored by the FAO, that had taken
place four years earlier. Participants in the tour had observed that a majority
of eucalypt species furnished firewood, charcoal, short poles, and mine props.
Some were good for timber and wood pulp, explained the Australian hosts,
and additional ones made excellent windbreaks or could be counted on to
help dry out wet sites.

Ninety-six participants from twenty-six nations attended the Rome conference. Twenty-nine were from Italian institutions, and seven came from
France, including the redoubtable Métro, who led a general discussion about
"problems basic to planting." Australia, the United States, Belgium, the United
Kingdom, and Portugal each supplied five or six representatives. Jacobs was a
member of the Australian delegation, and he gave a paper about the human
use of eucalypts in Australia.[10]

The U.S. section was chaired by Woodbridge Metcalf, professor of forestry at the University of California, Berkeley. Metcalf had inherited the eucalypt trials program at the University's Santa Monica station in southern California from Norman Ingham. Ingham had devoted a good part of his career
as a forester to researching the alien trees. Metcalf's paper, "Eucalyptus Trees
in the United States," was a useful summary of the work going back forty years
to his days in Santa Monica.[11]

Other countries supplied one or two delegates, occasionally more. They
included Thailand, Cuba, Brazil (4), Chile, Argentina, Iran (2), Sudan, Indonesia, and India (2). Curiously, nobody represented South Africa, although
British participants held posts in Nyasaland and Northern Rhodesia. Six technical papers presented at the meeting addressed the overall utility of eucalypts;

four described problems in planting; five discussed tree establishment, management, and protection; two were about protecting farmlands and soils; and nine focused on timber and industrial uses. In addition, twenty-four nations submitted country reports dealing with the status of eucalypts and the types of research and publications pertaining to them. Country reports from such disparate areas as Japan, Turkey, Ethiopia, and Haiti demonstrated the spread of the genus.

In the discussions and recommendations published in the conference proceedings, the conferees noted that "eucalypts can play a much more important part in [the] wood economy of all countries if these difficulties [in sawing, seasoning, and such] can be overcome." Problems needed to be tackled in a coordinated way, so that data and information could be shared; the FAO was designated as the conduit for data and publications.[12]

Participants were optimistic about the future of the genus. Métro lauded the steps that had been taken to include eucalypts in forestry planning and research, with study methods standardized by IUFRO. Participants agreed to share their findings and, in order to avoid duplicating research, asked existing regional forestry commissions, already under FAO auspices, to establish working parties on eucalypts. Governments were to set up teams of specialists, and it was resolved to hold a meeting every five years. The FAO was to arrange a second world conference in order to sustain the momentum already generated.[13]

The second conference took place in São Paulo in August 1961. Two hundred forty delegates from nineteen nations attended the five-day meeting, during which participants gave a total of 140 papers. This was more than double the number of participants at the Rome meeting. This Second World Conference on Eucalypts focused on the commerce and use of eucalypts in the tropics and paid special attention to the status and outlook for the genus in Latin America and Africa.[14]

Regional reports noted that 1.7 million ha (4 million acres) (800,000 ha [2 million acres] in Latin America, 500,000 ha [1.2 million acres] in Brazil, 400,000 ha [1 million acres] in Africa and Mediterranean Europe) of plantations existed outside Australia, a figure that seemed paltry in comparison with Australia's total of more than 4 million ha (11 million acres) of indigenous eucalypt forests. However, plantations were expanding rapidly beyond Australia, and their combined annual productivity exceeded that from natural stands.

Twelve species were supplying fuelwood, charcoal, posts, cellulose, and timber (in descending order of importance), and were assisting in the stabilization and reclamation of soils. The final conference report alluded to problems of identification, a perennial issue, and to the need to select high-quality seed stock. To this end Australian expertise figured prominently. Conferees welcomed the suggestion for a "eucalyptus assistance service" that would organize bulk shipments of seeds from Australia, and also the suggestion that Australian botanists should revise the classification of the genus, another long-standing concern. It was deemed proper to adopt Australian methods of determining wood density and shrinkage and to use that nation's standards for sawn wood and its methods for researching diseases.

Finally, participants agreed to contact the Australian government in regard to the possibility of a 1966 conference. However, the multifaceted character of research and the inclusion of eucalypts in a range of FAO-sponsored symposia on wood technology, tree breeding, and plantation forestry, anticipated the need for a third world conference on eucalypts. The World Symposium on Man-Made Forests, sponsored by the FAO, took place in Canberra in April 1967. Topics included conifers, teak, poplars, and other trees, as well as eucalyptus, and tours included gum trees in their native habitats, plus forests and plantations of other species in New Zealand and New Guinea. By that time, eucalypt forests totaled 1.7 million ha (4.2 million acres) worldwide.[15]

International gatherings, such as commonwealth forestry conferences, IUFRO world conferences, world forestry congresses, and a variety of symposia or seminars ("Man-Made Forests" [1967], "Tree Breeding" [1970, 1978, 1989], "Tree Improvement" [1985], "Forest Management" [1987, 1998], "Plantation Forestry" [1987], "Fast Growing Trees" [1989], and "Regional Forest Agreements" [1998]) have all at one time or another singled out eucalypts. The plant genus features most prominently in discussions about wood planting and production because of its ease of cultivation, adaptability, and rapidity of growth.[16]

A recent joint Australian-Japanese workshop, held in Canberra in October 1995, reinforced the significance of eucalypts in national and regional economies. Referring mostly to the humid tropics, participants recognized that eucalypts were "more effective in establishing plantations on degraded sites than species indigenous to such regions." Foresters discussed how best to supply high-quality seed and technical know-how in order to maximize wood products from these plantations. China figures prominently in such deliberations, and efforts are under way to boost growth by establishing Australian ectomycorrhizae, a soil fungus that makes nutrients more available to the roots of a compatible host, as cost-effective inoculations into nursery soils. Additional discussions at the 1995 meeting included planting hybrids on salt-affected and acid soils (*E. camaldulensis* in Pakistan) and using plantations (*E. viminalis* in Australia) as a way of recycling effluents and biosolids discharged by wastewater facilities.[17]

Organizations committed to "exotic forestry" have also collaborated through symposia and workshops to exchange data about cultivation methods and outlets. The Association Forêt-Cellulose (based in France), the Technical Association Pulp and Paper Industry (in the United States), and the IUFRO (in Italy) have actively included eucalypts in meetings about forest products research.[18]

The coordination of research at the international level has brought important issues to light, making comparisons possible between different nations. Eucalypts in tropical Africa, for example, have performed differently from their counterparts in Latin America. Part of the reason for the difference concerns the influence on plant performance of source regions for seeds. Foresters observed differences in survival among nine *E. grandis* provenances and five *E. saligna* provenances at high-altitude sites (for example, in Zimbabwe). In general, *E. saligna* did much better than *E. grandis*. Higher altitude *E. grandis* plants from Australia survived cold temperatures in Africa better than ones from lower sites in Australia, and intermediate ones from east-central Queensland did much better in Zimbabwe than counterparts imported from Brazil.[19]

Clearly, international organizations are well situated to coordinate studies about resilience to frost or drought, conduct breeding programs, and compare and analyze trials with different populations or species in a range of environments. Subsidies and support staff are able to relieve regional and national governments of research and development costs. With the results from planting programs, such agencies can inform large or small countries, rich or poor economies, about what uses and outlets exist for eucalypts.

The tier of expertise and specialized research, however, may also serve to distance foresters from the actual experience and ramifications of growing eucalypt plantations. Consultants usually do not live or work for very long in the vicinity of their favorite plants. Consequently, local anxiety and complaints about what the trees do to patterns of life and livelihood may or may not reach them. Experts can address the unexpected difficulties, or may be buffered from or inured to environmental or social fallout from new plantations. In this latter case, foresters are dealing with a constituency that may have little contact or empathy with local people. In this sense, international expertise, as expressed by contract assignments, ignores or treats as irrelevant those factors that do not fit into the framework of specialized forestry. The social costs and cultural reverberations of accepting "aid" in the form of eucalypts is thereby subordinated to the need to grow more and more wood, and to plant those "miracle trees" that can do it.

Conclusion

Give them a fair showing of place and climate, and they will thrive and enrich their
environment. This tree has the hardiness of
the ancient; it also has virtues which will
enlarge the comforts and lengthen the days
of men.

—SAMUEL LOCKWOOD, 1878

Recently, grass-roots protests about eucalyptus grown for industrial pulp and wood chips have broken out in Thailand. Farmers objected to government land being leased to rich industrialists, who set out plantations of *E. camaldulensis*. The use of national forest reserves for these enterprises has sparked the most opposition. The companies, often owned or financed by foreign capital, dispossess or intimidate squatters who live in the forest reserves. In 1990, 110,000 ha (272,000 acres) of fast-growing gum existed in Thailand, a ninefold increase in barely five years. Officials speculated that plantations would eventually cover 4.16 million ha (10.3 million acres)—a large portion of Thailand's so-called degraded forest land. Recently, that estimate has been revised downward to about 2.4 million ha (6 million acres). Whatever the final figure, it is evident that plantations will provide relatively few jobs and result in the eviction of thousands of poor families.[1]

Buddhist monk Phra Prajak Kuttajitto argues that the natural forest is more subtle and complex than officials or consultants are willing to admit. Increased runoff, downstream flooding, and the felling of native trees to make way for eucalypts offend forest dwellers, he notes. Prajak lives in the forest and has used a variety of tactics to thwart and embarrass forestry interests. For example, he has "ordained" thousands of hectares as "Children of Buddha," causing some cutters to refrain from chopping those indigenous woods.[2]

In 1992 the Thai government suspended its reforestation program, partly because of such protests and partly because of concerns about the long-term impacts of new pulp mills. Linkages among officials, business executives, multinational corporations, and development agencies have surfaced as Thailand has geared up for international wood and pulp markets. Speaking about that country, environmentalist Larry Lohmann explained that "to a technocracy trained to concentrate on growth in export income and chronically susceptible to the blandishments of big business, eucalyptus seems a perfect way of cashing in the 'underutilized resource' of 'degraded' National Reserve Forest land."[3]

Despite a logging ban in Thailand, timber companies continue to exploit native forests. Instances of officials turning a blind eye to illegal cutting, then giving companies access to reforestation schemes on "degraded lands," and the companies in turn planting those lands in eucalypt plantations have repeatedly come to light. A Thai environmental group calculates that in the thirty years before 1990 a total of 640,000 ha (1.6 million acres) was planted in pines and eucalypts, yet in 1989 alone more than 800,000 ha (2 million acres) of native woodland was cut down. The ban is a smokescreen for continued felling, the group declares, and logging has spilled over the sparsely protected borders of Burma, Cambodia, and Laos. Reforestation with eucalypts and similar fast growers is the supposed antidote to this massive loss of vegetation cover.[4]

One reason why the cutting ban is ineffective, notes Caroline Sargent of the International Institute for Environment and Development in London, is that the issue is not eucalypts but rather control of land. With the ban, loggers were enabled to defend their domain from encroachment by some of the 8 million squatters who eke a living by clearing land. At the same time agriculturists who have taken possession of cleared land are unwilling to give up space for forest agriculture, citing the evils of eucalypts. Sargent concludes that in Thailand "the battle is not about Eucalyptus, however. It is about power."[5]

The Thai situation continues to fester. In spring 1996 ten thousand or more farmers camped for almost a month near the prime minister's office in Bangkok. Among other issues, they were protesting what they saw as the excessive numbers of eucalypts. Rural people and a number of researchers have continued to express indignation about the area being planted in the alien trees and about the seemingly insatiable thirst that gum species manifest. Industrial pulp exporters fared poorly in the economic downturn of the late 1990s.[6]

How did the perception of eucalyptus change so drastically, from the unqualified support of early enthusiasts to thousands of protesting farmers besieging government offices for days on end in opposition to the genus? Put another way, why is there such a sharp contrast between the optimism and exuberance of von Mueller, Cooper, Kinney, Navarro de Andrade, and others, and the current widespread hostility toward the trees in parts of the world? It is surprising and disquieting to set the early, upbeat pronouncements about the benefits of eucalypts against the shrill invective and cynical commentary heaped upon the species and their custodians in recent years. What was intended to reward the efficient forester and thrifty citizen with shade, timber, firewood, medicines, industrial oils, and honey is now dismissed by many as a "green hell" or as a harsh and demanding "green desert." What went wrong? Did early promoters simply miscalculate? Were they misguided—so taken by their successes that they assumed nobody would ever object to their "miracle" trees? Were expectations built up so high that they were bound to be dashed, and promotion and adoption would inevitably lead to de-adoption? Is the problem one of scale, since like any monocrop, eucalyptus plantings tend to overwhelm local ecologies and affect traditional methods of making a living?

Three aspects of the introduction and spread of eucalyptus help answer these questions. First, it is clear that certain expectations and promises went unfulfilled. Second, there was an understandable faith in the testimony and arguments of expert scientists. Botanists, foresters, and horticulturists lent their professional clout to popularization of the plants. The data they published about tree growth and potential uses for the wood convinced many people of the merits of planting these novelties from Australia. Third, the pace and scale of planting have altered the regard local people have for the genus. Set out along roads, waterways, fence lines, and in scattered copses, eucalypts might tie into the contours of a landscape and the needs of inhabitants. But as man-made plantations for industrial timber have multiplied and expanded exponentially in size, eucalypts have come to dominate by their sheer extent and density. It was this shift toward industrial forestry that began to change public beliefs about the benefits of eucalyptus.

Expectations

One of the most important reasons for introducing eucalypts was to provide timber and firewood. Various national and international expositions and fairs trumpeted the merits of the plant genus. Engineers recommended larger specimens as pilings in docks and piers and as planks for shipbuilding. Smaller trees went for poles and agricultural implements. The broad range of uses in Australia, however, did not transfer readily to other nations. The hardness of the wood and its propensity to warp, crack, and shrink while immature disappointed commercial outlets. Some growers made money from poles, charcoal, and firewood, but not to the extent that they initially expected.

Disappointment in the end product and its marketability arose from two causes. First, as petroleum began to replace wood for heating, power generation, and transportation, particularly in North America, the outlook for eucalypts dimmed. In Brazil railroads incorporated eucalypts into their sources of firewood and used them as fencing for railroad tracks, but the same interests in California soured on the wood, especially after track ties were laid while the wood was green. Transport officials in California turned to other materials. There was also a natural reluctance on the part of both manufacturers and consumers to replace well-known woods with new and inadequately tested ones, such as eucalypts.

Disappointment also resulted from the decline of early plantings themselves. Some failures were due to ad hoc methods of obtaining seeds. Many shippers gathered seeds from convenient or accessible sites without regard for the habitat requirements of various species. Often seeds were obtained from a single tree, or a grove, irrespective of whether the specimens appeared healthy and vigorous. Additionally, mixing eucalypts on estates and in arboreta promoted hybridization and led to confusion and complaints about genetic variability and loss of vigor.[7]

Recently, Australian expert Ross Florence has concluded that a eucalypt's "land race" (a term denoting the place in which a species has grown for a long time) is not necessarily the best source for continued research and breeding. For example, the land race for *E. grandis* in South Africa grows more slowly than high-quality stock from near Coffs Harbour, New South Wales. In another trial, the growth rate of this species did not match that of trees from native areas in Australia. Similarly, *E. hybrid* has not grown as rapidly in recent tests in India as either *E. camaldulensis* or *E. tereticornis* (believed to be the source of original hybridization) carried in from north Australia. Species and subspecies that have grown in an area for a long time may not necessarily be the best suited to that location. Efforts to upgrade older stock are one way of bolstering a plant's sagging reputation in one or more foreign regions.[8]

As a "fever tree," eucalyptus failed to cure respiratory ailments generally associated with marshes and low-lying land. Touted as a remedy for bronchial problems and a variety of environmentally linked febrile disorders, eucalypts fell into disfavor because planting trees to cleanse the environment simply did not work. As the concept of the organic nature of disease replaced ideas rooted in ancient beliefs, eucalypts lost out. Although the new trees did draw off surface waters and helped dry swampy areas—"balancing" a surfeit of moisture that could lead to ill health, according to a traditional view—claims about their value in dispelling malaria by disinfecting the atmosphere and littering the ground with antiseptic leaves proved erroneous. Eucalypts assumed more mundane functions that other tree species also satisfied, such as drying soils, protecting crops from wind and frost, or lending shade over streets and parks.

Initial interest and enthusiasm exemplified by early research into tree selection, growth, wood properties, and site characteristics, as well as cultivation and management, supplied a large body of data from which a reenergized commercial and industrial sector was able to draw. Today, eucalypts are novelty plants in California. "Baby eucalypts" make decorous plants for homes, offices, and shopping malls. Specimens are kept trimmed at about chest height and last for ten or twelve years before being replaced. The aesthetic qualities of the genus, which interested early European growers, are carving out a multimillion-dollar market among florists and interior decorators throughout North America. In California, nurseries for these ornamentals are more profitable than for roses and orchids, and in Hawaii, 350,000 genetically engineered eucalypts are being planted on abandoned sugar plantations.[9]

Pulp, not ornament, is the major use for eucalypts today. The pulp market for eucalypts took off in the 1960s. At the first world eucalypt conference in Rome in 1956, discussions of that product were subordinated to other uses, such as firewood, charcoal, and windbreaks. However, within twenty years that position changed. Today, pulp dominates the debate about the value of the plant genus. Eucalyptus grown for pulp is especially lucrative for developing countries with relatively inexpensive land, few environmental regulations, low wages, inexpensive startup costs, and high returns in the form of exports

case when a native plant species has a restricted range and uses for it are conjectural.

Changes of Scale

The scale of eucalypt planting increased markedly after World War II. International agencies and postcolonial assistance programs geared up tree planting in order to raise living standards, especially in developing countries. One basic aim was to grow more firewood. Consequently, the dialogue about whether to plant eucalypts took place among scientists, planners, and officials. The relationship between expert and consumer grew more tenuous when international agencies and funding institutions promoted eucalypts for reforestation, drought mitigation, and land renewal without regard to local contexts.

Foresters and agency personnel do not merely respond to wood shortages; rather, their research improves wood production and quality, and thus opens new opportunities for the commercial wood market. Commercial pulp producers apply the new information and techniques to increase company profits and growth. Pulp industrialists and engineers may be more concerned about what bankers, administrators, and officials think than about the problems of local residents in some of the sites they target for plantations. This results in the scenarios mentioned in Thailand, India, and Portugal. Local people learn to regard the trees as symbols of authority or of trickle-down economics that ignores traditions of land use and livelihood.

Residents may also make their voices heard over eucalypts for purely political reasons. In modern, growth-oriented economies, such as Brazil, people are more sensitive to a power structure that has shut them out. Environmental groups may also fuel these political challenges. Activists inform local people about happenings in other places, pointing out how new eucalypt plantations will lead to disruptions. In this sense, eucalypts serve as a lightning rod for social dissent; they are scapegoats for issues of justice and social participation.

Eucalypts are not always anathema in local communities; indeed, they can be a great benefit. Ten villages encircle the Chingra Mouza forest in Midnapore, southwest Bengal, India. Intensive felling of native sal woodlands under forest department leases and for fuelwood resulted in major timber declines in the 1970s. In response, a local forest officer attempted to plant eucalypts, but he abandoned his effort after villagers cut the young trees for poles and fuel. Aware of the degraded conditions, a group of local men obtained eucalypt seedlings from the forester and planted them. Through hard work and political negotiation, including defiance of those who wanted to cut saplings, the Chingra men guarded their plant progeny. Within five years, nine of the ten villages around Chingra forest had begun to protect the vegetation cover.[12]

Revegetation resulted from cooperation between local villagers and forestry staff. Once community leaders recognized the benefits of replanting with eucalypts, they spearheaded overall plant recovery. Unfortunately, this case is all too rare in places like India, where national and state government authority is so entrenched. State forestry tended to replace natural forests with monocultural plantations, mainly eucalypts from 1960 to 1980. This phase resulted in severe ecological and social disruption. Rather than return to a "Ghandian" sense of local autonomy that eschews industrialization, it might be preferable to consolidate and promote links between tree farmers (like the Chingra example) and regional or industrial markets. This is the argument of two recent authors, M. Gadgil and R. Guha. Modified use of eucalypts fits into their schemes for afforestation and wood use.[13]

The enlarged scale of plantings and authoritarian support for eucalypts have formed a wedge between those charged with oversight, usually government officials and scientists, and those who live within the ambit of new forestry. What began as a novelty among a small cadre of plant lovers has evolved into a forest plantation industry whose scale, size, and economic importance could barely have been envisaged. Clearly, the opportunity to establish tropical plantations in developing nations with cheap land and little capital (exemplified by Latin America and, more recently, Southeast Asia) has resulted in the enormous spread of commercial forests. Short-rotation stands of a single species, selected to render a uniform product, are perfect in such schemes.[14]

Experts suggest that soon half of the world's industrial wood will be supplied by forest plantations. Much will come from the tropics where expansion occurred, notes forester Julian Evans, at a rate of about 2 million ha (5 million acres) annually in the 1980s, grown mostly on a short rotation for fuel and pulp. Eucalypts account for at least one-third of this total, and their significance will undoubtedly grow. Speculators believe that Latin America will continue to be a major area for plantation development, with Brazil and Chile dominating the eucapulp stream. Forecasters have no doubt that eucalypts, for pulp and for firewood, will continue to be a powerful incentive in both land clearance and land revegetation. Like it or not, gum trees are here to stay. Rather than contest the terrestrial area that is being devoted to them, it is better to establish a dialogue that focuses upon the right site for the right species for the right reasons, and to include in the purview the cultural practices and values of those people who will be affected by the introduction and establishment of eucalypts.[15]

Appendix 1.
Commonly Planted Eucalypts

Sources: Doran and Turnbull, eds., *Australian Trees and Shrubs*; Eldridge et al., *Eucalypt Domestication and Breeding*; Evans, *Plantation Forestry*; Florence, *Ecology and Silviculture of Eucalypt Forests*; Hillis and Brown, *Eucalypts for Wood Production*.

E. botryoides, Southern Mahogany

Restricted to a narrow coastal zone in New South Wales and eastern Victoria, southern mahogany grows on exposed sites with strong winds. The red wood saws well and is strong; it has done fairly well in California, Brazil, Italy, and New Zealand, where it is used for flooring, decks, and veneers. Southern mahogany hybridizes with *E. camaldulensis,* and resulting trees have gone into North Africa. Experiments in Australia, where it is rarely planted, show that it grows well on sewage water in Victoria.

E. camaldulensis, River Red Gum

A medium to large tree, river red gum is the most widely distributed of all eucalypts and one of the most variable in its native range, which covers twenty-five degrees of latitude in Australia, from 12°48′ to 38°15′S. This gum species does best on alluvial silts and sands in riparian lowlands with at least 200–600 mm (8–25 in), and up to 1,200 mm (50 in), of annual rainfall. River red gum is probably the most widely planted of all tree species in arid and semi-arid environments outside Australia. It grows rapidly on poor soils with low precipitation, but can cope with floods. It does not tolerate prolonged frost or calcareous soils. It has a range of uses, including wood (for shade, ornament, firewood, charcoal, poles, and posts) and honey. It coppices for up to five rotations and copes with drought, salinity, and light frosts. Initial cultivation took place in the Mediterranean basin from seeds collected from southern temperate stock in southern Australia. The northern tropical form does best in seasonally dry tropics. Its dense, dark-textured wood is less valuable for pulp (in Spain, Morocco, and Portugal), but has featured in charcoal production for smelting iron and steel. Expansion in South America and Southeast Asia in the 1980s has helped raise plantation area to about 1 million ha (2.47 million acres). Approximately 1 kg of seed will provide enough plants for a 100 ha (247 acre) plantation area, although river red gum is suited for vegetative propagation.

E. citriodora, Lemon-Scented Gum

Lemon-scented gum, a smooth-barked, straight-stemmed, vigorous coppicer, occurs in the undulating landscape of Queensland, Australia, between 17° and 26°S. It is a tropical species that exists on poor gravels, podsols, and laterites where there is 600–800 mm (25–32 in) of rainfall. Widely planted overseas, this gum has done best in central Africa, South America, and India. It is extensively planted for citronella oil, firewood, and poles.

E. deglupta, Kamarere

Kamarere is a tropical tree growing between 9° and 11°S, and is the only eucalypt species found naturally north of the equator. It does not occur in Australia. This species is fond of lowland riparian sites, consisting of fertile sandy loams, volcanic ash, and pumice. Pure stands will grow as high as 1,800 m (6,000 ft) above sea level in an annual moisture regime typically of 2,500–3,500 mm (100–140 in). Plantations of this species may replace tropical rainforests and currently total 100,000 ha (250,000 acres) in equatorial regions. Growth rates are very swift (up to 3 m/yr), but kamarere does not coppice. Wood and bark furnish short-rotation pulp. Fire is a danger to this species, and its sensitivity to site, disease, and pests (fungi, termites, and moths) makes this eucalypt difficult to grow. Most forest plantations are in South and Southeast Asia (Sri Lanka, Indonesia), Brazil (about 10,000 ha [25,000 acres]), Central America (Costa Rica, Cuba, Honduras), and Africa (Congo, Ghana, Ivory Coast), where there is a prolonged dry season and annual rainfall between 1,000 mm and 1,500 mm (40 and 60 in). Efforts are being made to expand cultivation in Malaysia, Fiji, and the Solomon Islands.

E. globulus, Southern or Tasmanian Blue Gum

One of the earliest species to be collected (1792) and described (1799), and the first to be cultivated outside Australia, southern blue gum ranges from 38°26′ to 43°30′S in Victoria, New South Wales, with concentrations in east Tasmania, from which many plantings in Europe originated. It grows from sea level to 1,100 m (3,600 ft) on granites, dolerites, and mudstones, being also partial to loams and well-drained heavy clays. In Tasmania, annual rainfall averages between 550 mm and 1,000 mm (22 and 40 in; at 500 m [1,650 ft] altitude at 43°S). Worldwide, the forest area of southern blue gum exceeded 800,000 ha (2 million acres) in the mid-1970s, more than for any other eucalypt. In recent decades it has expanded to about 1.3 million ha (3.2 million acres), but it is likely that *E. grandis* and probably *E. tereticornis* and *E. camaldulensis* have exceeded that amount. Forest plantations thrive in mild (not below –6°C), drought-free climates. Southern blue gum coppices well (on eight- to twelve-year rotations over three cycles) on a variety of soils and produces firewood, pulp, and timber of variable quality. This species adjusts to Mediterranean-type climates and to higher altitudes in the tropics, but does not thrive in temperate areas with severe winters, tropical lowlands, or regions with long, hot dry seasons. Provenance trials were initiated in Tasmania in 1976, supplemented by collections from New South Wales and Victoria, in order to upgrade land races in foreign nations where provenances remain obscure.

E. grandis, Flooded or Rose Gum

Widely distributed along Australia's east coast from 16° to 32°52′S, flooded gum exists from sea level to 1,100 m (3,600 ft) across a range of fertile soils (except flood-prone ground) where average annual precipitation is between 700 mm and 2,500 mm (30 and 100 in). In 1918 it was separated from *E. saligna,* a close relative, although there are no recorded hybrids. Flooded gum is a plantation favorite due to its ease of care and rapid growth, and the superior qualities of its wood, used for pulp, charcoal, poles, and light construction timber. The wood is typically pale, rather soft, and has a less dense texture than that of other eucalypts. Today, this species is probably the most widely planted industrial eucalypt, covering more than 2 million ha (5 million acres; mainly in Brazil and South Africa). It is subject to frost damage, although programs are under way to harden off pulpwoods. Breeding programs by Aracruz, Brazil, involved vegetative propagation. Interspecific crosses with *E. urophylla* help make pulps more fungus-resistant.

E. obliqua, Messmate

Messmate, the type species for the genus, is widespread and economically valuable, extending around coastal regions of southeast Australia and Tasmania, from 29° to 43°S. It adjusts to a range of edaphic conditions and can tolerate rainfall of 750–1,250 mm (30–50 in) per year. Different-sized trees furnish saw timber, fuelwood, and pulp in Australia, but plantations have not flourished abroad, primarily because messmate is sensitive to frost in cool, moist areas, where other eucalypts do better, and also because it grows more slowly than other widely used species. There are small plantations in New Zealand.

E. regnans, Mountain Ash

The second tallest tree in the world (after California's coastal redwoods), mountain ash has a discontinuous distribution in Victoria and Tasmania between 37° and 43°S. It grows from 200 to 1,100 m (650 to 3,600 ft) above sea level on moderately fertile, well-drained soils with average rainfall of more than 1,000 mm (40 in) annually. Specimens more than 95 m (312 ft) high tower in fog-shrouded, even-aged, pure stands, which will hybridize naturally with *E. obliqua* where ranges overlap. Plantation specimens grow quickly and put on large volumes of low-density, light-colored, even-grained wood; the wood is used for furniture and pulp. Outside Australia, the tallest tree (82 m [269 ft]) comes from Kenya, where annual wood increments averaged 56 m³/ha/yr. Planted at higher altitudes in India, Sri Lanka, and Zimbabwe, mountain ash has prospered in sites not subjected to drought, severe cold, or poor soils.

E. saligna, Sydney Blue Gum

Scattered along Australia's east coast from 24° to 36°S, Sydney blue gum overlaps with *E. grandis,* preferring less damp sites (with annual rainfall of 800–1,500 mm [30–60 in]) and higher slopes with moderate loams, where it tolerates frost better than *E. grandis.* Forest plantation use is similar to *E. grandis,* although Sydney blue gum coppices better and provides wood for general construction and pulp. It is grown extensively in New Zealand, Brazil, East Africa including Angola, and Hawaii.

E. tereticornis, Forest Red Gum

Closely related to *E. camaldulensis,* forest red gum is extensively distributed through thirty degrees of latitude, including southern Papua New Guinea and eastern Australia into Victoria, from 10° to 38°S. It grows up to 1,000 m (3,300 ft) above sea level in a rainfall belt of 500–1,500 mm (20–60 in) annually. It prefers light clays, gravels, and loams, and adjusts to deep, well-drained, fairly light soils outside Australia. One of the first eucalypt species to be exported, forest red gum has prospered in areas with summer rain and a dry season, as found in the tropics and subtropics of Africa and Asia, where an original seed source in India's Nandi Hills resulted in the "Mysore Hybrid." Mysore gum is the most widely planted eucalypt in India (about 500,000 ha [1.2 million acres]), especially in deforested and barren areas. Hybridization may involve *E. robusta* and *E. camaldulensis.* A single tree on Madagascar is reported to have been the seed source for wholesale plantings in West Africa and also in China. One or two trees in Zanzibar reportedly led to plantings in East Africa. Used for firewood, pulp, posts, charcoal, hardboard, and mine timbers, forest red gum has flourished in many nations, including Brazil (with 250,000 ha [618,000 acres]), Uruguay, Argentina, Colombia, Israel, Mozambique, Angola, Zaire, Ghana, Vietnam, and Sri Lanka. It proves tolerant of a long dry season, but trees from northern portions of its natural range are susceptible to frost, doing best in humid tropical savannas in which the forest red gum must compete with grasses and termites during early years.

E. urophylla, Timor Mountain Gum

Timor mountain gum stands 25–45 m (82–150 ft) tall and exists on seven islands of the Lesser Sunda group from 7°30′ to 10°S. It covers a 500 km east-west band, principally on tropical montane sites on Timor Alor and Wetar, where it grows on well-drained volcanic soils to above 2,000 m (6,600 ft) above sea level with 600–2,500 mm (25–100 in) of summer monsoon rain annually. Timor mountain gum does not occur in Australia. Early plantations in Brazil, named *E. alba,* were of this species and persisted for several generations (expanding to about 500,000 ha [1.24 million acres]) from trials at Rio Claro. Recent reinvigoration of forest plantations in the wet/dry tropics consists of careful selections from provenances expected to perform well. They include plantings in Brazil, Indonesia, and south China. In 1890 Dutch researchers introduced seeds into Java; more arrived in Brazil in 1919, when Navarro de Andrade obtained some from the Bogor botanical gardens. He subsequently distributed seeds of "Brazil alba" widely because of their resistance to stem canker.

E. viminalis, Manna Gum or Ribbon Gum

Widespread in southeast Australia, including Tasmania, manna gum adjusts to a broad range of physiographic conditions, including precipitation between 500 mm and 2,000 mm (20 and 80 in) annually. It thrives on moist, fertile slopes, but adapts to thinner soils and grows from sea level to 1,000 m (3,300 ft) between 29° and 43°S. It is useful in temperate forest plantations, combining rapid growth with frost tolerance (down to –10°C), and produces short-rotation pulp for paper and hardboard. It coppices well and withstands fire damage. Worldwide, its areal extent exceeds 200,000 ha (494,000 acres), with a nucleus in southern Africa. However, insect damage led to abandonment in the eastern Transvaal (until biological controls enhanced survival) and to losses in the Black Sea region due to cold weather (1949–50). Currently, manna gum does well in southern Brazil and Argentina, shows promise in Chile, Turkey, and Portugal, and is established in Italy, Spain, and India.

Appendix 2. Tables

TABLE 1
Area under Eucalypts (Mainly Plantations)
(in thousands of hectares)

Country	1970s	1990s	Country	1970s	1990s
Algeria	28.2	30.0	Laos	—	62.5
Angola	100.7	135.0	Lesotho	0.4	2.0
Argentina	80.0	236.0	Libya	26.0	26.0
Australia	26.4	75.0	Madagascar	135.0	130.0
Bangladesh	—	12.0	Malawi	9.6	30.0
Bolivia	5.0	—	Malaysia	7.0	8.0
Brazil	1,000.0	3,617.0	Mali	—	5.0
British Solomon			Mauritius	3.0	3.0
Islands	32.0	1.5	Mexico	2.0	38.0
Burkina Faso	—	7.0	Morocco	178.0	200.0
Burundi	18.6	40.0	Mozambique	8.0	14.0
Cameroon	2.0	13.0	Myanmar	0.6	40.0
Canary Islands	0.6	?	Nepal	—	6.0
Chad	0.2	1.5	New Zealand	12.7	22.0
Chile	31.1	180.0	Nicaragua	—	5.5
China	52.0	1,870.0[1]	Niger	—	2.0
Colombia	13.8	31.0	Nigeria	4.0	11.0
Congo	5.5	35.0	Pakistan	1.0	28.5
Costa Rica	—	10.0	Papua New Guinea	1.3	10.0
Cuba	36.0	35.0	Paraguay	2.9	8.0
Cyprus	2.0	?	Peru	92.9	211.0
Ecuador	17.7	44.0	Philippines	7.1	10.0
El Salvador	—	2.0	Portugal	250.0	500.0
Ethiopia	42.3	95.0	Puerto Rico	0.8	?
Gabon	—	2.0	Rwanda	23.0	60.0
Ghana	0.9	14.0	Senegal	0.1	40.0
Greece	0.1	?	S. Africa	347.5	538.0
Guatemala	—	6.0	Spain	390.3	350.0
Haiti	—	2.0	Sri Lanka	8.3	29.0
Honduras	0.3	0.5	Sudan	7.6	23.0
India	450.0	4,800.00[2]	Taiwan	—	3.5
Indonesia	—	80.0	Tanzania	2.7	25.0
Iran	3.0	?	Thailand	0.1	195.0[3]
Iraq	3.0	?	Togo	0.3	10.0
Israel	10.0	10.0	Tunisia	42.0	42.0
Italy	38.0	40.0	Turkey	10.8	20.0
Jamaica	—	0.08	Uganda	11.5	10.0
Kenya	11.3	17.0	United States	110.0	110.0

Appendix 2

TABLE 1
(continued)

Country	1970s	1990s	Country	1970s	1990s
Uruguay	111.1	160.0	Zambia	7.5	26.0
Venezuela	—	70.0	Zimbabwe	25.0	30.0
Vietnam	—	1,000.0			
Zaire	5.0	20.0	Total	3,855.8	15,577.6

SOURCES: Métro, *Eucalypts for Planting*; Jacobs, *Eucalypts for Planting*; Davidson, "Ecological Aspects," pp. 39–40; Vercoe, "Australian Trees on Tour," pp. 189–92; Midgley et al., "Exotic Plant Species," p. 3.

[1]Midgley et al., "Exotic Plant Species," p. 3, notes 670,000 ha (1.65 million acres) in plantations, plus 1.2 million ha (3 million acres) equivalent in community plantings.

[2]Midgley et al., "Exotic Plant Species," p. 3, lists 2.5 million ha (6.2 million acres) for India, whereas Vercoe, "Australian Trees on Tour," p. 189, lists 4.8 million ha (11.9 million acres).

[3]Midgley et al., "Exotic Plant Species," p. 3, lists 195,000 ha (482,000 acres) for Thailand, compared with Vercoe, "Australian Trees on Tour," p. 189, who lists 62,000 ha (153,000 acres).

Appendix 2

TABLE 2

Eucalypts outside Australia

Date	Country	Comments
1774	England	*E. obliqua* seed planted in Royal Botanic Gardens at Kew, London.
1790	Java	Dutch gathered seed of *E. urophylla* for planting in Bogor and Jakarta; 1800, *E. globulus* from Australia.
c. 1790	India	Several species planted in Nandi Hills of Mysore (Karnataka); 1843, first plantations of *E. globulus* in Nilgiri Hills of Madras Presidency (Tamil Nadu), as fuelwood. About 170 species and subspecies tested.
by 1800	Mauritius	Port of call for Australia fleets. Reportedly *E. robusta*. About 3,000 ha (7,400 acres) planted by 1970.
by 1803	Italy	Planted by Cesati at Camalduli, near Naples, in 1829, from seeds offered by Dehnhardt; plantings at Tre Fontane, Rome, in 1870, and widely diffused by 1950s with about 38,000 ha (94,000 acres) in 1975.
c. 1804	France	*E. globulus* seeds sent to Jardin des Plantes, Paris; 1810, in Malmaison; 1813, in Toulon; late 1850s, Ramel's promotion; about 1,100 ha (2,700 acres) in 1961.
1816	Russia	Crimea plantings failed; 1860, Black Sea Coast; 2,300 ha (5,700 acres) in 1957.
1823	Chile	*E. globulus* planted between Valparaiso and Arauco, intended originally for Peru. About 31,000 ha (77,000 acres) by 1966.
1825	Brazil	*E. robusta, E. tereticornis* in Rio de Janeiro Botanic Gardens. Plantations after 1900.
1828	S. Africa	Seedlings from Mauritius perhaps as early as 1807; after 1850, plantations of *E. globulus, E. tereticornis,* and *E. camaldulensis*; 347,000 ha (857,000 acres) by 1973.
1829	Portugal	*E. globulus* into Oporto; extensive plantings after about 1875, totaled 250,000 ha (618,000 acres) by 1974. A pulp mill started in 1923.
1843	Pakistan	Plantations for fuelwood by 1867.
1847	Spain	Botanic Gardens of Madrid distributed trees to several provinces. A pulp mill operated in 1949, and the planted area totaled 390,000 ha (964,000 acres) by 1973.
1853	Uruguay	Seeds from S. Africa to Botanic Gardens in Montevideo; 111,000 ha (274,000 acres) by 1973.
1853	United States	California and Hawaii (with 82 species) and 250 for United States as a whole.
1854	Algeria	*E. globulus* seeds from Prosper Ramel in France; 28,000 ha (69,000 acres) by 1965.
1857	Argentina	*E. globulus* on D. H. Bunge's land in Buenos Aires Province. Area planted totaled 80,000 ha (198,000 acres) in 1973.
1860	Peru	*E. globulus* in the Sierras, with 93,000 ha (230,000 acres) by 1975.
1862	Greece	State planted around Athens.
1865	Ecuador	*E. globulus* reportedly introduced by Garcia Moreno; about 18,000 ha (45,000 acres) planted by 1975.
1865	Egypt	Planted by Gastinal-Bey in Cairo's Botanic Gardens.
c. 1869	Trinidad	*E. tereticornis* in Botanic Gardens.
1872	Indonesia	*E. globulus* added to indigenous species.
c. 1875	Japan	Foreign missionaries and teachers made unsuccessful plantings around Tokyo.

TABLE 2
(continued)

Date	Country	Comments
1876	Singapore	*E. citriodora* and *E. globulus* to Botanic Gardens from Australia.
c. 1878	Cyprus	Planted by the British; 68 species by 1934.
c. 1880	Sri Lanka	First plantations about a decade later, extending over 8,000 ha (20,000 acres) by 1974.
1883	Malaysia	Seedlings of *E. robusta* from Queensland, Australia.
pre-1884	Tanzania	Possible plantings on coast by Arab or Indian traders.
1884	Israel	Australian *E. camaldulensis* seed for marsh reclamation around Mikve: 1,800 ha (4,450 acres) by 1955.
1885	Turkey	*E. camaldulensis*; large-scale afforestation in 1950s to 11,000 ha (27,000 acres) by 1974.
c. 1890	Zimbabwe	12 species, led by *E. globulus* by 1965 with 25,000 ha (62,000 acres).
1890	China	Initially from Italy to Canton, Foochow and Hong Kong. By 1990 more than 200 species introduced.
1895	Ethiopia	Emperor Menelik II planted *E. globulus* for fuelwood near Addis Ababa. By 1974, 42,000 ha (104,000 acres) planted.
end 1800s	Nepal	Ornamentals for Kathmandu (British Embassy) from India.
end 1800s	Angola	After 1949 Benguela Railway Co. established plantations, covering about 100,000 ha (247,000 acres) by 1970.
end 1800s	Madagascar	Plantings by missionaries, settlers, and foresters. About 180,000 ha (445,000 acres) planted by 1974.
1900	Bolivia	*E. globulus* seed from Argentina, covering about 5,000 ha (12,400 acres) in 1973.
1900	Crete	Planted by first governor Prince George.
c. 1900	Guatemala	*E. citriodora* for oil.
c. 1900	Sudan	*E. hybrid* from India.
c. 1900	Morocco	*E. camaldulensis,* 1918; *E. gomphocephala,* 1920; about 25,000 ha (62,000 acres) in 1955. Mill established in 1957, and 178,000 ha (440,000 acres) by 1974.
1902	Kenya	Trials for 19 species for fuel, especially *E. globulus* and *E. saligna.* Plantations totaled 11,000 ha (27,000 acres) in 1973.
1904	Vietnam	Single tree at Coc Leu, N. Vietnam.
1905	Thailand	French established in colonial gardens.
1908	Mexico	1950, trials begun for 27 species.
1911	Libya	Additions of *E. camaldulensis* from Sicily, with 26,000 ha (64,000 acres) in 1965.
1912	Uganda	*E. grandis, E. tereticornis.*
1920s	Lesotho	Town ornamentals and windbreaks.
1922	Myanmar	Seven species planted in first 5 years.
1929	Puerto Rico	*E. robusta.*
1930	Cameroon	*E. maidenii, E. robusta* and *E. saligna* best.
1931	Burundi	Seed from Tanzania, Zimbabwe, and S. Africa; at least 19,000 ha (47,000 acres) by 1974.
1935	Bali	*E. urophylla.*
1935	Upper Volta	*E. camaldulensis,* with 900 ha (2,224 acres) by 1976.
c. 1940	Paraguay	About 10 species acclimated by 1955.
c. 1950	Nigeria	Trials of 4 species in Bamenda province; 4,000 ha (9,900 acres) by 1973 with more than 50 species tried.
c. 1950	Zaire	Plantations in east and southeast with 22 species by 1955, and covered about 5,000 ha (12,350 acres) a decade later.

TABLE 2
(continued)

Date	Country	Comments
1952	Bandung	*E. urophylla*.
1953	El Salvador	*E. deglupta* and 3 others; 18,000 ha (44,500 acres) in 1974.
1956	Congo	*E. tereticornis* seed from Madagascar, and *E. urophylla*.
1964	Honduras	*E. robusta* from Brazil, and *E. grandis* from Brazil and Australia; with about 300 ha (741 acres) in 1972.
1965	Costa Rica	Trials in 1968 with 3 species.
1965	Yemen	27 species introduced at El-Kod.
by 1970	Rwanda	About 23,000 ha (57,000 acres) planted.
1971	Papua New Guinea	*E. urophylla*.
by 1973	Sierra Leone	Trials.
by 1973	Ivory Coast	Experimental plantings of 3 species.
by 1973	Ghana	About 900 ha (2,224 acres) planted.
by 1973	Dominican Republic	Trials and ornamentals.
by 1973	Kuwait	Ornamentals.
by 1973	Panama	Trials.
by 1973	Somalia	Trials.
by 1973	Central African Republic	Trials.
by 1973	Iran	Trials.
by 1973	Mali	Trials.
by 1973	Belize	Trials.
by 1973	Malawi	About 10,000 ha (25,000 acres) planted.
1979	French Guyana	*E. urophylla*.

SOURCES: Métro, *Eucalypts for Planting*; Penfold and Willis, *The Eucalypts*; Zacharin, *Emigrant Eucalypts*; Jacobs, *Eucalypts for Planting*; Davidson, "Ecological Aspects of *Eucalyptus* Plantations"; Stier, "World Eucapulp Industry."

Appendix 2

TABLE 3

Chronology of Aracruz Celulose S.A. (Brazil)

1967   First eucalypts planted by Aracruz Florestal using seeds of *E. grandis, E. saligna,* and *E. alba* (Brazilian form) from sources in the state of São Paulo.

1970   *E. alba* discontinued due to poor yields and uncontrolled hybridization over many generations.

1972   Canker disease leads to new, resistant seeds being imported: *E. grandis* from Zimbabwe, and *E. urophylla* from Timor, plus technical assistance from Australian tree improvement experts.

1975   Pulp mill designed. Large-scale vegetative propagation decided upon in order to meet economic goals, with assistance from French and Australian researchers.

1978   First mill of Aracruz Celulose (established in 1972) begins operations with a capacity of producing 500,000 tons per year (tpy) of eucalyptus pulp.

1979   The factory makes 291,000 tons of pulp and exports 286,000 tons at $520/ton through its Atlantic Ocean port of Barra do Riacho (renamed Portocel in 1985). Eucalypt cloning initiated.

1980   More than 100 million eucalypts (representing over 50 species) are growing on approximately 59,000 ha (145,800 acres) of company land.

1981   The president of Brazil, accompanied by four ministers and the state governor, opens electrochemical units that produce sodium chlorate and chlorine soda, adding to self-sufficiency in chemical use. Total pulp production tops 400,000 tons, of which 291,000 tons is exported, with a net profit of $41.5 million. The company controls 77,670 ha (191,922 acres) in three municipalities in Espirito Santo, of which 58,300 ha (144,059 acres) is planted with eucalypts.

1984   Aracruz scientists receive the Marcus Wallenberg Prize for research into eucalypt propagation by vegetative means. They studied 54 species, 1,254 provenances, and 5,000 selections from established plantations.

1988   Ground-breaking begins for a new bleach kraft eucalyptus pulp mill. The company voluntarily restricts the area under eucalypts to 132,000 ha (326,200 acres).

1990   Erling S. Lorentzen, chairman of the board, signs the Business Charter for Sustainable Development drafted by the International Chamber of Commerce. Company-owned area tops 203,000 ha (502,000 acres). A program called "Partners in Timber" is launched with local farmers, who cultivate seedlings supplied by Aracruz for household use and sale to the company.

1991   President of Brazil opens a second unit, Mill B, with capacity of 525,000 tpy. Overall output for Aracruz expands to 1.025 million tpy. Aracruz receives the Liceu Technology Prize for sustainable development as applied to forests from the São Paulo Liceu of Arts.

1992   Aracruz is recognized by the Business Council for Sustainable Development at the U.N. Rio Conference on Environment and Development. More than 100 government officials, business leaders, and others, together with the cruise ship *Gaia Viking,* visit the plant during the conference. Portocel, busiest pulp terminal in the world, handles 104 vessels, exports 1 million tons of pulp cargo. Aracruz becomes first Brazilian pulp producer to be listed on the New York Stock Exchange.

1993   The Partners in Timber program signs 2,019 contracts in 43 municipalities, totaling 19,200 ha (47,400 acres). In November the program is suspended due to a lawsuit brought by the Brazilian Environmental Agency and the state of Espirito Santo. The labor force falls by 23% due to a worldwide drop in pulp prices ($318/ton as compared with $763/ton in 1989). A shake-up takes place among top executives.

1995   Sales exceed 1 million tons (at $875/ton), and plans are launched to diversify operations as a hedge against pulp price fluctuations. Focus is on eucalypt clones that produce lumber, plywood, and household goods.

TABLE 3
(continued)

1998   Production exceeds 1.2 million tons, and with a $300 million modernization program completed is projected to reach 1.7 million tpy. Pulp costs remain low, yields increase (41 m³/ha/yr) on 138,000 ha (341,000 acres), with more than 100 clonal species (mainly hybrids of *E. grandis* and *E. urophylla*). Luiz Kaufmann, CEO and president, resigns, and Cabs Aquiar, who began as a process engineer in 1981, succeeds him.

1999   Start-up of first sawmill—largest hardwood mill in Brazil—at a cost of $40 million in order to produce high-quality eucalyptus woods.

SOURCES: Aracruz Celulose S.A., *Annual Reports*, 1980–94; Marcus Wallenberg Foundation, *The New Eucalypt Forest*; *Gazeta Mercantil*, 26 March 1998; and Cody, "Aracruz Continues."

Notes

Preface

1. See Puntasen, "Political Economy of Eucalyptus"; Boland, "Australian Trees in Five Overseas Countries," p. 76; Environmental Defense Fund, "Failure of Social Forestry in Karnataka"; Montalbano, "'Fascist' Trees."

2. Penfold and Willis, *The Eucalypts*, p. xix. Zobel, Van Wyk, and Stahl, *Growing Exotic Forests*, pp. 35, 43, claim that *E. camaldulensis* with *Pinus radiata* is the most widely planted exotic forest tree; *E. grandis* runs a close second.

Chapter 1 / The Geography of Eucalypts

1. Florence, *Ecology and Silviculture of Eucalypt Forests*, p. 15, suggests that there may be as many as a thousand species and forms, which occur in all but very dry areas. See A. G. Brown and W. E. Hillis, "General Introduction," in Hillis and Brown, eds., *Eucalypts for Wood Production*, p. 3; Pryor, *Biology of Eucalypts*, pp. 1–2; and Hillis, "Fast Growing Eucalypts," pp. 184–93. No eucalypts occur naturally in New Caledonia or New Zealand, although members of the Myrtaceae family exist on these islands. New Zealand has achieved technical breakthroughs with introduced eucalypts. *E. deglupta* was a species increasingly planted in West Africa for the purpose of growing wood pulp.

2. Boland et al., *Forest Trees of Australia*, pp. 193–96; Florence, *Ecology and Silviculture of Eucalypt Forests*, pp. 1–3; Australian Bureau of Agricultural and Resource Economics, *Forest Products Statistics*, p. 50.

3. Cremer et al., "Stand Establishment," pp. 81–135; Pryor, *Biology of Eucalypts*, pp. 2–3; Cromer, "Role of Eucalypt Plantations in Timber Supply and Forest Conservation in Australia," pp. 299–310.

4. English names are taken from Hillis and Brown, eds., *Eucalypts for Wood Production*, app. 1, pp. 405–6.

5. *E. regnans* belongs to the subgenus *Monocalyptus*, which has one cap of a flower bud, which falls off as the flowers open, rather than two. This subgenus is called *Eucalyptus, sensu stricto*, because about a hundred species, including this one, are identified from descriptions based on the originally named species *E. obliqua*. See Eldridge et al., *Eucalypt Domestication and Breeding*, pp. 18–20. Florence, *Ecology and Silvicul-*

ture of Eucalypt Forests, p. 35, notes *E. regnans* as the most demanding of all the ash-group species in terms of biophysical needs. It requires at least 1,200 mm of rainfall per year, grows best on well-drained but moist soils, endures water stress badly, and must be sheltered from hot and cold winds.

6. Turnbull and Pryor, "Choice of Species and Seed Sources," p. 34. Early claims were repeated by von Mueller, *Select Extra-Tropical Plants,* p. 131, who noted surveyor G. W. Robinson's assessment of 471 ft (144 m) for *E. regnans* on Mount Baw-Baw.

7. Florence, *Ecology and Silviculture of Eucalypt Forests,* p. 40.

8. Turnbull and Pryor, "Choice of Species and Seed Sources," pp. 17–21; Métro, *Eucalypts for Planting,* pp. 412–13. Eldridge et al., *Eucalypt Domestication and Breeding,* pp. 92–93, note that the tallest specimens in Europe top 70 m (230 ft) and come from Portugal at 40°N latitude.

9. See Scarascia-Mugnozza et al., "Freezing Mechanisms, Acclimation Processes and Cold Injury in *Eucalyptus* Species," pp. 81–94. I have seen a 12 m specimen in a garden in Beverley, East Yorkshire, that is at least thirty years old and that supported a colony of corvids called rooks. These birds fashioned bulky stick nests in its upper branches until the owner shot them.

10. Turnbull and Pryor, "Choice of Species and Seed Sources," p. 57; Moulds, "Eucalypts and Their Use in Semi-Tropical Plantings," p. 10.

11. Pryor, *Biology of Eucalypts,* p. 5; Eldridge et al., *Eucalypt Domestication and Breeding,* pp. 4–5, 24; Davidson, "Ecological Aspects of *Eucalyptus* Plantations," p. 38.

12. Eldridge et al., *Eucalypt Domestication and Breeding,* 23, Florence, *Ecology and Silviculture of Eucalypt Forests,* 47–50 (quotation p. 47). Pryor and Johnson, "Eucalyptus: The Universal Australian," note that eucalypts "have been a phenomenally successful group in Australia since the mid-Tertiary" (p. 533). Flannery, *The Future Eaters,* pp. 217–36.

13. McIlroy, "Grazing Animals," pp. 139–42; Carne and Taylor, "Insect Pests"; Hillis, "Fast Growing Eucalypts," p. 185; Pryor, *Biology of Eucalypts,* pp. 72–76; Keast et al., *Birds of Eucalyptus Forests and Woodlands*; Smith, *Saving a Continent,* p. 54.

14. Jacobs, "Eucalyptus as an Exotic," p. 5.

15. Scarascia-Mugnozza et al., "Freezing Mechanisms, Acclimation Processes and Cold Injury in *Eucalyptus* Species."

16. Eldridge et al., *Eucalypt Domestication and Breeding,* pp. 45–46.

17. Ibid., pp. 46–47.

18. Hillis, "Chemicals," pp. 357–60. Métro, *Eucalypts for Planting,* p. 146, notes that *E. astringens* is the basis for a tannin industry in Australia.

19. Baker and Smith, *Research on the Eucalypts and Their Essential Oils,* pp. 458–59; Boland, "Brief History," pp. 3–10.

20. The Medical Advisory Service for Travellers Abroad (MASTA) recommends "mosiguard" made from eucalyptus oil as a gentler alternative to insect repellent DEET (diethyl toluamide); see Blair, "Deadly Buzz"; Illman, "Happiness Is Needle-Shaped"; Doran, "Commercial Sources."

21. Métro, *Eucalypts for Planting*, pp. 31–32; D'Ombrain, *Gallery of Gum Trees*, p. 7; Moulds, "Eucalypts and Their Use in Semi-Tropical Plantings," p. 9; Kinney, *Eucalyptus*, p. 147 (quotation); C. Hall, *On Eucalyptus Oils*.

22. See Lawless, *Encyclopedia of Essential Oils*, p. 93; S. Price, *Aromatherapy Workbook*, p. 80.

23. Métro, *Eucalypts for Planting*, pp. 146–47, relies on Blakely's work in assessing the quality of gum-tree honey.

24. Ingham, *Eucalyptus in California*, pp. 110–11.

25. Moulds, "Eucalypts and Their Use in Semi-Tropical Plantings," pp. 2–3; Grimwade, *Anthology of Eucalypts*, p. 8; Penfold and Willis, *The Eucalypts*, pp. 1–8.

26. Bentham and von Mueller, *Flora Australiensis*, 3:185: "shrubs or trees, attaining sometimes a gigantic size, secreting more or less resinous gums, whence their common appellation of *Gum-trees*." See also Zacharin, *Emigrant Eucalypts*, p. 51.

27. Kinney, *Eucalyptus*, p. 42 (quotation).

28. Maiden, *Critical Review*, 1:1–11; Grimwade, *Anthology of Eucalypts*, p. 8; Pryor and Johnson, *Classification of Eucalypts*; Florence, *Ecology and Silviculture of Eucalypt Forests*, pp. 4–6.

Chapter 2 / Bound for Europe: Eucalypts for Pleasure

1. Zacharin, *Emigrant Eucalypts*, pp. 43 and 47 (quotation); Moyal, *Bright and Savage Land*, pp. 9–13; Beaglehole, ed., *The Endeavour Journal of Joseph Banks*, 2:5 (quotation). Brooker and Kleinig, *Field Guide to Eucalypts*, 1:3, suggest it was *E. gummifera*; others, *E. alba*.

2. Beaglehole, ed., *The Endeavour Journal of Joseph Banks*, 2:66 (quotation), 71, 115, and pl. 22; Maiden, *Sir Joseph Banks*, p. 18. This species was planted commercially in large numbers in California during the first decade of the 1900s because it grew extremely swiftly and appeared to resist frosts; see Ingham, *Eucalyptus in California*, p. 74.

3. Penfold and Willis, *The Eucalypts*, p. xix; Baker and Smith, *Research on the Eucalypts*, p. 3. The full title of L'Héritier's work was *Sertum Anglicum, seu plantae rariores quae in hortis juxta Londinium imprimis in horto regio Kewensi excoluntur, ab anno 1786–87 observatae.*

4. Jacobs, *Genus Eucalyptus in World Forestry*, p. 2.

5. Penfold and Willis, *The Eucalypts*, p. xix; Boland et al., *Eucalyptus Seed*, p. 1; Beaglehole, ed., *The Endeavour Journal of Joseph Banks*, 2:115; Zacharin, *Emigrant Eucalypts*, pp. 39–41.

6. Kelly, *Eucalypts*, 1:1.

7. These estimates are derived from Métro, *Eucalypts for Planting*, p. 35, who estimated that eucalyptus grew on 700,000 ha throughout the world; Jacobs, *Eucalypts for Planting*, p. v; Evans, *Plantation Forestry*, pp. 35–36, tables 3.2 and 3.3. Evans esti-

mated that the total area in forest plantations in 1990 was 42.7 million ha, of which 37.5 percent was in eucalypts. His estimate for the beginning of the decade was 7.865 million ha (p. 35). Ball, "Development of Eucalyptus Plantations," p. 24, gives a similar total of 43.9 million ha in tropical forest plantations in 1990, and 17.8 million ha for 1980. Davidson, "Ecological Aspects of *Eucalyptus* Plantations," pp. 39–40, suggests that there were 13.41 million ha under eucalypt plantations in 1993.

8. Montalbano, "'Fascist' Trees"; Cromer, "Role of Eucalypt Plantations in Timber Supply and Forest Conservation in Australia"; Davidson, "Ecological Aspects of *Eucalyptus* Plantations," p. 40; Calder et al., eds., *Growth and Water Use of Forest Plantations,* p. xi.

9. This is the case for *E. globulus* and *E. grandis* in the mid-latitudes and *E. tereticornis, E. camaldulensis,* and *E. urophylla* and its hybrids in the tropics. See Brown and Hillis, "General Introduction," in Hillis and Brown, eds., *Eucalypts for Wood Production,* p. 3. Australia's plantations are mostly conifers, notably *Pinus radiata;* see Dargavel and Semple, eds., *Prospects for Australian Forest Plantations.*

10. Fletcher, *Story of the Royal Horticultural Society,* pp. 6–9, 136.

11. Loudon, *Arboretum et Fruticetum Britannicum,* pp. 958–59.

12. Ibid.; Loudon, *Hortus Britannicus,* part 1493, "Eucalyptus." Smith authored the thirty-six-volume *Sowerby's English Botany* (1790–1814) and *The English Flora* (1824–28). He was founder of the Linnean Society, and after he had purchased Linnaeus's entire library in 1783, he wrote several papers on Australian plants, which appeared in the *Transactions* of the Linnean Society (vol. 3, 1797); see N. Hall, *Botanists,* p. 119.

13. Loudon, *Hortus Britannicus,* part 1493, "Eucalyptus." There are no dates for seventeen species.

14. "The Eucalyptus," *Gardener's Magazine* 6 (1830): 502.

15. *Gardener's Magazine* 13 (May 1837): 23; 11 (November 1835): 570.

16. *Gardener's Magazine* 11 (March 1835): 151; 14 (October 1838): 453.

17. *Gardener's Magazine* 14 (November 1838): 511; 13 (October 1837): 463.

18. *Gardener's Magazine* 6 (1830): 502; *The Garden,* 26 February 1876, p. 200; Elwes and Henry, *Trees of Great Britain and Ireland,* 6:1627.

19. Robinson, *Subtropical Garden,* p. 134.

20. "Vegetation of the Island of St. Léger," p. 509.

21. C. T. Kingzett, *Nature's Hygiene* (London: Bailliere, Tindale and Cox, 1907), p. 363.

22. "Eucalyptus for English Gardens," p. 192; Vilmorin, "Flowers of the French Riviera," pp. 80–104.

23. See *Gardener's Chronicle,* 9 December 1843, p. 862, in which the Isle of Wight trees are named as swamp mahogany (*E. robusta*) and one from Italy as *E. multiflora.* See also *Gardener's Chronicle,* 30 March 1844, p. 199; 31 May 1851, p. 342; 21 June 1851, p. 390.

24. Landsborough, "Eucalypts in Scotland," p. 516; Elwes and Henry, *Trees of Great Britain and Ireland,* 6:1618, 1642–44. Zacharin, *Emigrant Eucalypts,* p. 61, suggests

that the initial introduction to Whittinghame may have resulted from James Maitland Balfour's visit to Tasmania in 1846.

25. *The Garden,* 27 May 1876, p. 491.

26. *Gardener's Chronicle,* 10 May 1851, pp. 291–92; 3 May 1851, p. 277.

27. "Jarrah Timber," pp. 188–89; "Eucalyptus Timber for Street Paving," pp. 219–21. Boland et al., eds., *Forest Trees of Australia,* pp. 258, 358, note that the tallest specimen of karri felled at Pemberton in 1901 was 87 m (285 ft).

28. Maiden, *Bibliography,* pp. 8–11. Von Mueller published a note about timber in Victoria for the Intercolonial Exhibition of Australasia in Melbourne in 1866.

29. Wood, ed., *Reports on the Colonial Section of the Exhibition,* pp. 421, 438.

30. "Blue Gum," p. 1.

31. Kingzett, *Nature's Hygiene,* pp. 354–77, takes note of many scientific papers during the 1870s. See also R. Bentley, "On the Characters," lecture to Fellows of the Royal Botanic Society of London, 14 March 1874, cited by Thompson, "Australian Fever Tree," p. 235; Bosisto, "Is the Eucalyptus a Fever-Destroying Tree?" pp. 10–23.

32. MacArthur is spelled Macarthen by Kingzett, *Nature's Hygiene,* p. 360, and Naudin, "Mémoires sur les Eucalyptus," p. 379.

33. N. Hall, *Botanists,* pp. 45, 87, 100; Naudin, *Description et emploi des Eucalyptus,* p. 33.

34. Von Mueller, *Eucalyptographia,* 6:n.p. [p. 16]. Planchon, Eucalyptus Globulus *from a Botanic, Economic and Medical Point of View,* p. 19, lists Tristany's outlet as *Compilador Medico,* confirming popular regard for the trees on the Mediterranean coast of Spain. Adolphe Brunel of Toulon followed up with experiments, but died unexpectedly in 1871, leaving his heirs (and at least ten other researchers named by Planchon) to publish results, confirming the "anti-febrile properties of the new medicament beyond doubt" (p. 19).

35. The extract from *Nature* appears in Thompson, "Australian Fever Tree," p. 237n.24; *The Garden,* 12 October 1872, p. 310; Kingzett, *Nature's Hygiene,* p. 371.

36. O'Connor, "Cultivation of 'Eucalyptus Globulus,'" p. 133.

37. Brockway, *Science and Colonial Expansion,* pp. 108–15.

38. Kingzett, *Nature's Hygiene,* p. 371; "Eucalypts in Scotland," p. 522 (quotations); von Mueller, *Select Extra-Tropical Plants,* pp. 131, 136 (quotations). *E. amygdalina* (white peppermint) was the one species the author claimed as possessing the highest percentage of volatile oil of the eucalypts that had been tested.

39. Pepper, "Eucalyptus in Algeria and Tunisia," pp. 39–56. Pepper was a physician who was located in Algiers and had been a member of the society for ten years when he presented the paper. Von Mueller published an essay on eucalypts in *Bulletin de la Société Agricole d'Algérie* 11 (1868): 179–81, and 12 (1869): 106–17. Planchon, *Eucalyptus Globulus from a Botanic, Economic and Medical Point of View,* pp. 13–14, congratulates "distinguished colonists" M. A. Cordier and M. Trottier for ambitious efforts in planting trees in Algeria in the 1860s.

40. *The Garden*, 19 February 1876, p. 190; 27 May 1876, p. 492; 29 September 1877, p. 313. Von Mueller, *Eucalyptographia*, 6:15, corresponded with "Colonel" Playfair about the blue gum, which thrived in droughty conditions where "scarcely a grain of wheat could be reaped," transforming sterile lands with its twigs and foliage.

41. Planchon, Eucalyptus Globulus *from a Botanic, Economic and Medical Point of View*, p. 15.

42. Pepper, "Eucalyptus in Algeria and Tunisia," p. 45.

43. *The Garden*, 29 September 1877, p. 313; Kingzett, *Nature's Hygiene*, p. 366.

44. *The Garden*, 19 February 1876, p. 190; *Gardener's Chronicle*, 7 March 1874, p. 205.

45. *The Garden*, 2 May 1874, p. 386; 21 March 1874, p. 238.

46. *The Garden*, 30 October 1875, p. 374.

47. Chown, "Raising Fever Gum Trees," p. 434; Inchbald, "*Eucalyptus Globulus* on the Riviera," p. 202; Planchon, Eucalyptus Globulus *from a Botanic, Economic and Medical Point of View*, p. 15.

48. *The Garden*, 8 July 1876, p. 36; 16 November 1878, p. 441.

49. *The Garden*, 2 November 1878, p. 408; "On Ornamental Trees and Shrubs in the Garden of Castlewellan," p. 414; Elwes and Henry, *Trees of Great Britain and Ireland*, 6:1627.

50. *The Garden*, 4 June 1881, p. 568.

51. *The Garden*, 7 August 1880, p. 126; 6 May 1882, p. 317. N. Hall, *Botanists*, p. 129, notes that the firm of Vilmorin Andrieux & Compagnie was one of the world's premier seed houses, and that Henri Vilmorin served as president of the Botanic Society of France (1899) and visited Russia, the United States, and England on behalf of plant improvement and applied botany.

52. *The Garden*, 27 April 1889, p. 383; "Eucalypts in Scotland," pp. 515–31; Zacharin, *Emigrant Eucalypts*, pp. 60–63; Elwes and Henry, *Trees of Great Britain and Ireland*, 6:1619.

53. Zacharin, *Emigrant Eucalypts*, p. 65.

54. Planchon, Eucalyptus Globulus *from a Botanic, Economic and Medical Point of View*, pp. 9, 10 (quotations); Zacharin, *Emigrant Eucalypts*, pp. 51, 53; Duval, *King's Garden*, pp. 112–13, 123. Maiden, *Critical Review*, 2:249, quotes from "Voyage in Search of La Pérouse." See also Stafleu and Cowan, *Taxonomic Literature*, 2:710–11.

55. Zacharin, *Emigrant Eucalypts*, p. 56. The one species Bonpland mentions, *E. diversifolia* (coastal mallee), is undistinguished and has leaf differences (unknown to that early botanist) in its juvenile and adult stages, hence the Latin name.

56. P. Ramel, "L'*Eucalyptus Globulus*," p. 787; Kinney, *Eucalyptus*, p. 10, citing Planchon. Von Mueller, *Eucalyptographia*, 6:6, recalled that Ramel witnessed the rapid growth of blue gum in the Melbourne gardens from 1855 to 1857, though never in its "forest-haunts." He gives the date of Ramel's return to France as 1858.

57. Naudin, "Mémoire sur les Eucalyptus," pp. 337, 378, 379.

58. Zacharin, *Emigrant Eucalypts*, p. 68; Booker and Kleinig, *Field Guide*, p. 135; Elwes and Henry, *Trees of Great Britain and Ireland*, 6:1626. Eldridge, *Eucalyptus Camaldulensis*, pp. 2–3, notes that as recently as 1967 the more descriptive term was preferred by *Forestry Abstracts*.

59. Elwes and Henry, *Trees of Great Britain and Ireland*, 6:1626.

Chapter 3 / Beyond Australia: Ferdinand von Mueller

1. Chisholm, *Ferdinand von Mueller*, pp. 4, 16–17; von Mueller, "On the General Introduction of Useful Plants into Victoria," p. 99.

2. Details about *Pinus radiata* are supplied by Clapp, "Unnatural History," pp. 1–19; Lavery and Mead, "*Pinus radiata*," pp. 432–49.

3. Von Mueller, "Anniversary Address," p. 3; von Mueller, "On the General Introduction of Useful Plants into Victoria," pp. 93–105. I have relied upon Moore, "Green Gold," for an excellent discussion of von Mueller's character, scientific significance, and legacy. Kynaston, *Man on the Edge*, p. 1, refers to von Mueller as a "Victorian living in a Victorian society in a Victorian state" with all the pretense and platitudes the term implies.

4. Dücker, "Australian Phycology," pp. 123–24; Kynaston, "Exploration as Escape."

5. Churchill, Muir, and Sinkora, "Published Works of Mueller," pp. 10–11.

6. Willis, *By Their Fruits*, pp. 10–13; Dücker, "Australian Phycology," pp. 123–24. Willis, Dücker, and others note that William Hooker wrote to the governor on behalf of von Mueller and secured him the appointment. Hooker does not seem to have noted von Mueller's earliest publications, which appeared in German periodicals. See Cohn, "Ferdinand Mueller, Government Botanist"; Maroske and Cohn, "'Such Ingenious Birds'"; Lucas, "Baron von Mueller."

7. Chisholm, *Ferdinand von Mueller*, p. 3; Willis, *By Their Fruits*, p. 18.

8. Dücker, "Australian Phycology," p. 123.

9. *Australian Dictionary of Biography*, vol. 5; Churchill, Muir, and Sinkora, "Published Works of Mueller," pp. 11–12. The number of publications, awards, and so forth is subject to confusion: see the explanation by Churchill, Muir, and Sinkora, and also the list supplied by Moore, "Green Gold," p. 23.

10. Von Mueller, *Eucalyptographia*, 10:n.p. [p. 15].

11. Cooper, *Forest Culture*, p. 61; von Mueller, *Select Extra-Tropical Plants*, pp. 131–32, 141.

12. Bosisto, *Indigenous Vegetation of Australia*, p. 5; also N. Hall, *Botanists*, p. 24.

13. Bosisto, *Indigenous Vegetation*, pp. 5, 6. *Heysen's Gum Trees* reproduces nine watercolors of farm and riverine landscapes, including "Murray Cliffs," which is dominated by red-toned gums, sandstone cliffs, and cattle; and "A Grey Morning," showing blue-colored gums in a dawn mist. Other painters, such as John Glover, Louis Buvelot, E. Phillips Fox, and William Ford, also painted eucalypts.

14. Von Mueller, *Eucalyptographia*, 10:n.p. [pp. 8–9].

15. Von Mueller, *Select Extra-Tropical Plants*, pp. 132–35.

16. McClatchie, *Eucalypts Cultivated in the United States*, p. 3.

17. Von Mueller, *Eucalyptographia*, 1:n.p. [p. 6].

18. "William Saunders," p. 625.

19. Kynaston, *Man on the Edge*, pp. 29, 42. Sara Maroske, personal communication, 25 February 1997, suggests that in some years von Mueller penned as many as six thousand or more letters, all hand-written.

20. Willis, *By Their Fruits*, p. 181; Gildas, "L'Eucalyptus dans la campagne romaine," p. 180. Von Mueller, *Eucalyptographia*, 10:n.p. [p. 7], refers to "his Grace Dr. J. A. Goold, R. C. Archbishop of Melbourne," and to the species of eucalypts as blue gums.

21. Gildas, "L'Eucalyptus dans la campagne romaine," pp. 180–85.

22. U.S. State Department, Consular Report 168, pp. 7, 13.

23. Kingzett, *Nature's Hygiene*, pp. 368–69.

24. See "Procès-Verbaux."

25. Von Mueller, *Eucalyptographia*, 10:n.p. [p. 16].

26. Von Mueller, "Forest Culture in Its Relations to Industrial Pursuits," in Cooper, *Forest Culture*, pp. 63–65. This discussion and statistics appear almost verbatim in Drury, *Useful Plants of India*, pp. 200–202, attesting to the authority of von Mueller's pronouncements.

27. Von Mueller, who died in 1896, is buried in St. Kilda Cemetery, on Dandenong Road in Melbourne.

Chapter 4 / California Promotes Eucalypts

1. Details about the first introductions into California differ. Santos, *Eucalyptus of California*, pp. 16–18, lists several individuals likely to have pioneered plantings, declaring that Walker was "involved early in the propagation of eucalyptus in California" (p. 16). He relies in part on forester Woodbridge Metcalf for information about growers in the East Bay and discusses the possible role of retired clipper-ship captain Robert Waterman, who planted eucalypts in the Suisun district of the North Bay in 1853. Warren, "Eucalyptus Crusade," also offers suggestions about pioneers, as does Tyrrell, *True Gardens of the Gods*. Kinney, *Eucalyptus*, reported that C. L. Reimer successfully introduced fourteen species in January 1856; Warren, p. 32, notes that that number existed in Walker's nursery.

2. See Woodson, "Eucalyptus Boom and Bust"; Sellers, *Eucalyptus*, p. 9; Thompson, "Australian Fever Tree," pp. 234–35; Santos, *Eucalyptus of California*, p. 18.

3. Santos, *Eucalyptus of California*, p. 19.

4. Santos, *Eucalyptus of California*, pp. 21–22. See also Groenendaal, "California's First Fuel Crisis," pp. 129–36, who notes that Captain Robert H. Waterman arranged to have gum tree seeds imported in 1853, and set them on a large tract of land

he had purchased. Waterman gave seeds to settlers of Fairfield, a community he founded. Santos (p. 21) notes that another sea captain, Joseph Aram, opened a nursery in San Jose and set out blue gums along the Milpitas Road in 1856.

5. Groenendaal, "California's First Fuel Crisis," pp. 120–21, 135–40; Santos, *Eucalyptus of California*, p. 21.

6. Cat Spring, *Century of Agricultural Progress*, p. 67.

7. Cook, "Distribution of Certain Species," pp. 5–40; Arbenz, *Eucalyptus in Texas*.

8. Cooper, *Forest Culture*, touted as "the only complete and reliable work on the Eucalypti published in the United States." See also Warren, "Eucalyptus Crusade," p. 33; N. Hall, *Botanists*, p. 40.

9. McClatchie, *Eucalypts Cultivated in the United States*, p. 18. Santos, *Eucalyptus of California*, p. 23, uses the word "milestone."

10. Warren, "Eucalyptus Crusade," pp. 34–35; Woodson, "Eucalyptus Boom and Bust"; Thompson, "Australian Fever Tree," p. 235n.19, citing the 1874 report of the commissioner of agriculture. Santos, *Eucalyptus of California*, p. 22, mentions Cooper and J. L. Barker, also from Santa Barbara County, planting the "first large acreage in southern California," and refers (p. 26) to Sutro's promotion in San Francisco.

11. McClatchie, *Eucalypts Cultivated in the United States*, p. 19.

12. McClatchie, *Eucalypts Cultivated in the United States*, pp. 19, 22; N. Hall, *Botanists*, p. 79.

13. McClatchie, *Eucalypts Cultivated in the United States*, p. 19; Santos, *Eucalyptus of California*, p. 26; Warren, "Eucalyptus Crusade," p. 38; Kinney, *Eucalyptus*, p. 118. See also Kinney, "Eucalyptus," in Fernow, ed., *Some Foreign Trees*, pp. 23–28, for contacts he established with government foresters.

14. See "Crisis: Eucalypts," p. 6.

15. Woodson, "Eucalyptus Boom and Bust," p. 76; Metcalf, "Eucalyptus Trees," pp. 91–96; Santos, *Eucalyptus of California*, p. 31.

16. McClatchie, *Eucalypts Cultivated in the United States*, p. 31; N. Hall, *Botanists*, p. 88.

17. Kinney, *Eucalyptus*, p. 13.

18. McClatchie, *Eucalypts Cultivated in the United States*, p. 31.

19. Details about eucalypts appeared in U.S. Forest Service, *Circular* 59 (1907), 179 (1910), 210 (1912); U.S. Forest Service, *Bulletin* 196 (1908), 87 (1911); California State Board of Forestry, *Bulletin* 1 (1910); California State Board of Forestry, *Circular* 2 (1907); California Agricultural Experiment Station, *Bulletin* 196 (1908), 225 (1911). During that era H. M. Hall's article, "Key for Identification of California Eucalypts," appeared in the *Encyclopedia of American Horticulture,* and other useful references appeared in *The Hardwood Record* (1913), *Journal of Forestry* (1921), and the *Proceedings* of the Society of American Foresters (1907).

20. McClatchie, *Eucalypts Cultivated in the United States*, pp. 31–32 (quotation p. 31).

21. Ibid., p. 34. A note appeared in the *Los Angeles Times*, 29 April 1901, that four hundred fruit growers from ten farm clubs in Cucamonga had learned, from information presented at a horticultural gathering, about air temperatures remaining higher on the north side of a windbreak where trees shielded citrus from damaging temperatures.

22. McClatchie, *Eucalypts Cultivated in the United States*, pp. 31, 33–35. For use as windbreaks, see Metcalf, "Eucalyptus Trees," pp. 91–96.

23. McClatchie, *Eucalypts Cultivated in the United States*, pp. 35–36; Metcalf, *Growth of Eucalyptus*, p. 43.

24. McClatchie, *Eucalypts Cultivated in the United States*, p. 36; Kinney, *Eucalyptus*, p. 39.

25. Metcalf, *Growth of Eucalyptus*, pp. 42–43.

26. Ibid., pp. 42, 45, 47. Red gum *(E. rostrata)*, gray gum *(E. tereticornis)*, manna gum *(E. viminalis)*, plus sixty to seventy-five additional species growing in the state, although not in plantations, p. 25. Ken Eldridge, personal comunication, March 1997, notes that only in the 1990s did Australian and other research into wood sawing and seasoning establish ways to process sawn products profitably from young, fast-grown eucalypts.

27. McClatchie, *Eucalypts Cultivated in the United States*, p. 38.

28. Metcalf, *Growth of Eucalyptus*, pp. 39–45.

29. Kinney, *Eucalyptus*, pp. 120–24; McClatchie, *Eucalypts Cultivated in the United States*, pp. 41–42, 88.

30. McClatchie, *Eucalypts Cultivated in the United States*, pp. 39–41.

31. Metcalf, *Growth of Eucalyptus*, pp. 44–45.

32. McClatchie, *Eucalypts Cultivated in the United States*, p. 42; U.S. State Department, Consular Report 168.

33. McClatchie, *Eucalypts Cultivated in the United States*, pp. 43–44.

34. Kinney, *Eucalyptus*, p. 149.

35. Ibid., p. 157.

36. Ibid., p. 162.

37. Ibid., pp. 159–60 (quotation on 160).

38. Ibid., pp. 132, 138, 139, and 141.

39. Ingham, *Eucalyptus in California*, p. 30.

40. Ibid., p. 31.

41. Ibid.; Metcalf, *Growth of Eucalyptus*, pp. 42–43; Eldridge, personal communication, March 1997.

42. Ingham, *Eucalyptus in California*, p. 77. The area in eucalypts is discussed by Thompson, "Australian Fever Tree," p. 244n.80.

43. Warren, "Eucalyptus Crusade," pp. 39–40; Woodson, "Eucalyptus Boom and Bust," pp. 78–79. On alternatives to using wooden ties for railroads, see U.S.D.A., Bureau of Forestry, *Bulletin* 3 (1889), 4 (1890), and 9 (1894).

44. Betts and Smith, "Utilization of California Eucalypts," pp. 5, 8.

45. Ibid., pp. 8, 14, 19–20.

46. Ibid., pp. 24, 26.

47. Ibid., pp. 26, 27.

48. Ibid., pp. 28–30. U.S. acting secretary of agriculture W. M. Hays approved the insert "Eucalypts in Australia" on 22 July 1910.

49. Kinney, *Eucalyptus*, pp. 35–36.

50. Pratt, *Use of Lumber*, p. 116.

51. Groenendaal, "California's First Fuel Crisis," p. 210.

52. Little, "Fifty Trees from Foreign Lands." The three species were *E. globulus, camaldulensis,* and *sideroxylon,* which would grow in a "hardiness zone" from central California southward into southwestern Arizona, along the lower reaches of the Rio Grande in Texas, and eastward along the Gulf of Mexico into southeast Georgia and Florida (p. 817). See also Stoeckeler and Williams, "Windbreaks and Shelterbelts."

53. King and Krugman, *Tests of 36 Eucalyptus Species.*

54. Recent wildfires in suburban Oakland and in Los Angeles highlight the risks from growing such inflammable vegetation in dense patches and stands. See "The Oakland/Berkeley Hills Fire," Firewise, 1992, available on-line at http://www.firewise .org/pubs/The Oakland Berkeley Hills Fire/the-fire.html.

Chapter 5 / Industrialists and Eucalypts: Take-off in Brazil

1. Navarro de Andrade, "Eucalyptus in Brazil," pp. 215–20, 240 (quotation on p. 215); Boland, "Galloping Gums."

2. Navarro de Andrade, *O Eucalipto,* p. 6, dates the Rio Grande do Sul planting to 1868. See also Flinta, "Value of Eucalypts," pp. 76–83; Boland, "Galloping Gums," p. 31. Jacobs, *Forestry Development and Research, Brazil,* p. 3, noted that an *E. robusta* specimen planted in Rio in 1825 was producing viable seed in 1972.

3. Navarro de Andrade, "Eucalyptus in Brazil," p. 215.

4. *Australian Dictionary of Biography,* 10:381–83, notes that the London-born Maiden, who emigrated to Sydney in 1880, took over von Mueller's mantle by revising the *Eucalyptus* genus, popularizing uses for it, and publishing more than two hundred articles on Australian plants.

5. Navarro de Andrade, "Eucalyptus in Brazil," p. 216; Boland, "Galloping Gums," p. 32, citing a forester relative, Navarro Sampaio, in 1952.

6. Navarro de Andrade, "Eucalyptus in Brazil," p. 217; Boland, "Galloping Gums," p. 32.

7. Navarro de Andrade, "Eucalyptus in Brazil," p. 217.

8. Ibid., p. 219.

9. Ibid., pp. 219–20.

10. Zon and Sparhawk, *Forest Resources of the World,* 2:692–93, 709.

11. "Dr. Navarro de Andrade's Acceptance," p. 213.

12. Boland, "Australian Trees in Five Overseas Countries," pp. 47–48; A. Aubreville, "Quelques problèmes forestiers du Brésil," p. 114.

13. Couto and Betters, "Short-Rotation Eucalypt Plantations," detail the incentives and fallout of the forestry laws between 1967 and 1988. The basic disadvantage of the 1966 law was that a company had to pay taxes, then finance reforestation in order to get a rebate, whereas the 1970 law liberalized the situation by allowing anticipated taxes to be written off under the costs of replanting. Evans, *Plantation Forestry*, pp. 72–73, gives a total of 7.1 million ha (17.5 million acres) for Brazil's plantations in 1990.

14. Flinta, "Value of Eucalypts," pp. 76–83, gives an overview of the areal spread and use of the genus for South America. See also Jacobs, *Forestry Development and Research, Brazil*, pp. 4–9.

15. FAO, *Second World Eucalyptus Conference*, 1:61, 2:4.

16. FAO, *Second World Eucalyptus Conference*, 1:166. Other nations growing eucalypts included Argentina (Sydney blue gum, river red gum, forest red gum, and manna gum); Chile and Peru (blue gum); and Cuba (Sydney blue gum). For a recent discussion of charcoal, see Fung, *Eucalypts in Brazil for Charcoal*.

17. FAO, *Second World Eucalyptus Conference*, 1:62, 165–66.

18. Golfari, Caser, and Moura, *Zoneamento ecologico*; Pryor, Chandler, and Clarke, *Establishment of Eucalyptus Plantations*, pp. 2–4. Zobel, Van Wyk, and Stahl, *Growing Exotic Forests*, dedicated their work to Golfari.

19. Boland, "Australian Trees in Five Overseas Countries," p. 25; Davidson, "Ecological Aspects of *Eucalyptus* Plantations," pp. 39–40.

Chapter 6 / Eucalypts for Pulp

1. Lewington, *Plants for People*, pp. 54–55; Holusha, "Pulp Mills"; "Assessment of Alternative Pulping and Bleaching Technologies."

2. Lewington, *Plants for People*, p. 200.

3. Algar, *Forestry and Forest Products*, p. 41; Higgins, "Pulp and Paper." Recent cold soda pulping in Tasmania used *E. regnans* and *E. delegatensis* wood chips impregnated with a sodium hydroxide solution at 41°C for two hours. Samples from additional species were tested for basic density, brightness, tensile strength, tear index, and bulk. See Banham, Orme, and Russell, "Pulpwood Qualities."

4. FAO, *Pulping and Paper-Making Properties*, p. 392; FAO, *Second World Eucalyptus Conference*, 1:80; Teicher, "Manufacture of Newsprint"; Phillips, "Manufacture of Pulp and Paper."

5. FAO, *Pulping and Paper-Making Properties*, pp. 79, 97, 117, 127, 173, 211; Eldridge et al., *Eucalypt Domestication and Breeding*, pp. 131–32.

6. Dean, "Objectives for Wood Fibre Quality and Uniformity."

7. Lewington, *Plants for People*, p. 56; Boland et al., *Forest Trees of Australia*, p. 362.

8. Carrere, "Pulping the South," p. 206; Stefan, "Market Pulp Producers," p. 87; Swann, "Market Pulp Powerhouse," p. 22.

9. Swann, "Market Pulp Powerhouse," notes that ABECEL consists of Aracruz Celulose, Bahia Sul Celulose, Cenibra-Celulose, Nipo-Brasileira, Cia Florestal Monte Dourado, and Riocell, with a combined capacity of 2.26 million tons per year. See also Aracruz Celulose S.A., *Annual Report 1980*, p. 3; ABECEL, *Statistical Report*, p. 12; Carrere, "Pulping the South," p. 207; Sharman, "High-yielding Pulp," p. 37; [Brazil's Pulp Output], p. 41.

10. Details about wood shortages in Brazil are supplied by de Miranda Bastos, "Importance of Eucalypts"; Zon and Sparhawk, *Forest Resources of the World*, 2:692–719.

11. Evans, *Plantation Forestry*, pp. 72–73; Carrere, "Pulping the South," p. 208. Cody, "Brazil's No. 3 Eucalyptus Producer," notes that Votorantim Celulose e Papel, one of Brazil's fastest growing pulp-and-paper companies, continues to invest heavily in eucalypt pulp production for specialty paper grades. In 1998 two mills were turning out 2,000 metric tons per day, of which 800 tons went for export.

12. Argentine Republic, Ministerio de Económica, *Argentina*, pp. 3–9, states that plantations were the best of three choices, the other two being use of mixed tropical species or of fibers such as bagasse. Vale, "Production Goals," pp. 3–4, explores the issue of wood deficits in the 1960s.

13. Evans, *Plantation Forestry*, pp. 72–73. For information about the startup of Aracruz, see Aracruz, *Annual Report 1980*, and *Annual Report 1983*, p. 5.

14. See Stier, "World Eucapulp Industry"; Aracruz, *Annual Report 1980*; Cody, "Aracruz Continues," p. 55.

15. Aracruz, *Annual Report 1980*, and *Annual Report 1981*, p. 1.

16. See Jacobs, *Eucalypts for Planting*, pp. 422–32; Burgess, "Natural Occurrence"; Eldridge et al., *Eucalypt Domestication and Breeding*, pp. 103–13.

17. See Jacobs, *Eucalypts for Planting*, pp. 422–32; Aracruz, *Annual Report 1994*.

18. See Jacobs, *Eucalypts for Planting*, pp. 503–6. Jacobs, *Forestry Development and Research, Brazil*, p. 9, notes that Navarro de Andrade "finally gave the location of the provenance as Dutch Timor at close to latitude 10°S." See also Eldridge et al., *Eucalypt Domestication and Breeding*, pp. 144–53.

19. Aracruz, *Annual Report 1984*, p. 9, and *Annual Report 1989*. Other prize winners were Leopoldo Garcia Brandao, Yara K. Ikemori, and Ney M. Santos (awarded posthumously). In his speech before the Wallenberg Symposium, forester Bruce Zobel called the teamwork "a catalyst for a new and revolutionary method in forest regeneration," and said that "it has been effective throughout the world." Zobel, [Address to Aracruz Researchers], p. 1.

20. Schreuder, De Barros, and Hill, "Global Supply and Cost Price Structure"; Aracruz, *Annual Report 1994*; Hillis, "Fast Growing Eucalypts," p. 187. There has been an increase of 85 percent in pulp per ha, from 6 tons per ha per year (28 m³/ha/yr) in forests raised from seeds in the late 1960s through the late 1970s to 11 tons per ha (45 m³/ha/yr) in 1986–94.

21. ABECEL, *Statistical Report,* p. 1; Stier, "World Eucapulp Industry," p. 159.

22. Aracruz, *Annual Report 1993*; Singh, "Pulping Horizons," p. 31.

23. Schreuder, De Barros, and Hill, "Global Supply and Cost Price Structure," pp. 81, 88; Cavalcanti, "Brazil's New Forests," p. 16.

24. Foster, "Business and the Environment." Kaufmann resigned as CEO in 1998 and Cabs Aquiar replaced him.

25. "This Isn't Pulp Fiction," p. 25; "Producers Foresee Increase"; Gatti, "Environment—Latam"; Ball, "Development of Eucalyptus," p. 15.

26. Swann, "Market Pulp Powerhouse," p. 22. Carrere, "Pulping the South," n. 5, states that annual U.S. paper consumption is 332 kg (732 lb) per capita. Schreuder, De Barros, and Hill, "Global Supply and Cost Price Structure," p. 85, note that the 1960 production total may be an underestimate.

27. Griffith, "Reversal of Fortune," p. 65; Weeks, "When Will It End?" p. 49; Wyman, "Spending Roller-Coaster," p. 28; "Colcura," pp. 5–6; Higgs, "Going for Growth," p. 45. Stefan, "Market Pulp Producers," p. 87, notes that the Chilean firm is Compania Manufacturera de Papeles y Cartones (CMPC).

28. Stefan, "Market Pulp Producers," p. 87; Picornell, "Why Do Developing Countries Go into Pulp?" pp. 3–5; "World Pulp Prices."

29. Aracruz, *Annual Report 1993*; Rocha, "Brazilian Court." The *Irish Times,* 7 May 1993, reported that two environmental activists had been killed in Brazil, and the Bahia *Municipios,* 13 September 1994, reported thefts of pulp-sized trees in southern Bahia and northern Espirito Santo.

30. Rocha, "Brazilian Court."

31. "Sustainable Development," p. 29; Aracruz, *Annual Report 1994,* pp. 31–39.

32. Stefan, "Market Pulp Producers," p. 87; Sharman, "High-Yielding Pulp," p. 37; Holusha, "Pulp Mills."

33. Carrere, "Pulping the South"; Carrere and Lohmann, *Pulping the South,* pp. 152–55 (quotation on 152).

34. Carrere, "Pulping the South," notes activities of the National Indian Foundation (FUNAI) in Brazil and the government decree 1775/96 that paralyzed Indian claims. The *Gazeta Mercantil,* 26 March 1998, reported that King Gustav of Sweden had postponed a visit to Aracruz due to the conflict about Indian land claims, and that a Dutch missionary-engineer, who had been working with the Indians since 1995, had been arrested and was facing deportation.

35. Sutton, Pearson, and O'Brian, "World Paper Production," p. 58; Davidson, "Ecological Aspects of *Eucalyptus* Plantations," p. 40.

36. Evans, *Plantation Forestry,* pp. 33–36; Davidson, "Ecological Aspects of *Eucalyptus* Plantations," p. 38. Sharma et al., "World Forests in Perspective," p. 19, note that all plantation forests account for 3 percent of the world's forested area, although in Chile, Kenya, South Africa, China, United Kingdom, Ireland, and Spain, plantation forests produce significant quantities of industrial woods.

37. Urry, "World Pulp and Paper Industry"; Waghorn, "The Debate Goes On," pp. 32–37.

Chapter 7 / Colonial Adoption: Eucalypts for India

1. Rajan, *Versatile Eucalyptus*, pp. 3, 10, states that "according to one version" (p. 10) seedlings were presented to Tippu by French allies. Seeds came either from Mauritius or directly from Australia. Other sources suggest that Dutch merchants introduced eucalypts. D. N. Tewari, *Monograph on Eucalyptus*, p. 5, notes 1790 as the year of first introduction and states that about a total of about 360,000 ha (890,000 acres) of agro-forestry programs in eighteen states was planted to this introduced genus by 1992.

2. Joshi, *Troup's The Silviculture of Indian Trees*, 5:162–63; Kumar, "Eucalyptus in the Forestry Scene of India," 2:1106.

3. O'Connor, "Cultivation of 'Eucalyptus Globulus,'" pp. 120, 132.

4. Ibid., pp. 121, 124, 126. Rajan, *Versatile Eucalyptus* p. 3, increases the number of species and varieties to about 170. Archivist Sara Maroske, personal communication, 25 February 1997, has alerted me to Denison's knowledge of von Mueller's work. See *Australian Dictionary of Biography*, 4:46–53.

5. See Poffenberger and McGean, eds., *Village Voices, Forest Choices*, pp. 86–100, on the significance of German forester Deitrich Brandis.

6. Ghate, *Forest Policy*, p. 37; Sunder, "India," pp. 23–26.

7. Ibid., pp. 37–38, 40; Bhattacharya, *Social Forestry*, pp. 1–5.

8. Joshi, *Troup's The Silviculture of Indian Trees*, 5:248–59. Dwivedi, *Forestry in India*, p. 195, estimates 430,000 ha (1 million acres); Eldridge et al., *Eucalypt Domestication and Breeding*, p. 141, enlarge that total. Tewari, *Monograph on Eucalyptus*, pp. 39–40. Rajan, *Versatile Eucalyptus*, pp. 10–15, supplies details about early cultivation in Karnataka.

9. See Redhead and Anandarajah, "Planting of Eucalyptus on Tea Estates," for Sri Lanka.

10. Mathur, "Eucalyptus," 1:21; Joshi, *Troup's The Silviculture of Indian Trees*, 5:163.

11. Joshi, *Troup's The Silviculture of Indian Trees*, 5:163.

12. See Mathur, "Eucalyptus," 1:21.

13. Joshi, *Troup's The Silviculture of Indian Trees*, 5:163.

14. Ibid., 5:163–64.

15. Ibid., 5:164.

16. Ibid.

17. Ibid., 5:164–68; Rajan, *Versatile Eucalyptus*, p. 14; Boland, "Eucalypt Seed for Indian Plantations."

18. Joshi, *Troup's The Silviculture of Indian Trees*, 5:168–69. Rajan, *Versatile Eucalyptus*, p. 3, places the state areas for eucalypts differently; for instance, he lists Kar-

nataka as leading with 104,547 ha (258,127 acres) in 1983 (p. 4), whereas Tewari, *Monograph on Eucalyptus,* lists that state (Mysore) at 71,000 ha (175,000 acres). Tewari's area under eucalypts (p. 5) is close to Joshi's data, with the exception of Tamil Nadu with 29,000 ha (71,700 acres).

19. Mathur, Sagar, and Ansari, "Economics of *Eucalyptus* Plantations"; see also Saxena, *India's Eucalyptus Craze.*

20. Livernash, "India"; Malhotra and Sharma, "Need for Defining Wasteland," pp. 3–9.

21. The term *social forestry* is attributed to Jack Westoby, who in 1968 defined it as "forestry which aims at producing a flow of protection and recreation benefits for the community." See Bhattacharya, *Social Forestry,* p. 14; Sunder, "India"; Ravindranath et al., "Ecological Stabilization," pp. 293–94.

22. Eldridge et al., *Eucalypt Domestication and Breeding,* p. 141; Tewari, *Monograph on Eucalyptus,* pp. 39–40. Rajan, *Versatile Eucalyptus,* pp. 1–3, credits Mysore's chief conservator of forests, M. A. Muthanna, with first identifying Mysore gum as *E. chickaballapur,* after its provenance in the Chickaballapur range of Kolar district, close to the Nandi Hills. Later he called it *E. hybrid;* outside the state of Mysore it became the so-called Mysore hybrid.

23. The total plantation area is taken from Eldridge et al., *Eucalypt Domestication and Breeding,* and specifics for the states from Joshi, *Troup's The Silviculture of Indian Trees.* They are regarded as minimum estimates. For a review of the taxonomy, see Boland, "Eucalyptus Seed," p. 131.

24. Joshi, *Troup's The Silviculture of Indian Trees,* 5:237–38; Dwivedi *Forestry in India,* p. 195; Tewari, *Monograph on Eucalyptus,* p. 6; Rajan, *Versatile Eucalyptus,* p. 2.

25. Joshi, *Troup's The Silviculture of Indian Trees,* 5:231–37.

26. Malhotra and Sharma, "Need for Defining Wasteland"; Boland, "Eucalyptus Seed," p. 120.

27. Shiva and Bandyopadhyay, "Eucalyptus."

28. Ibid., p. 187.

29. Ummayya and Dogra, "Planting Trees," p. 186.

30. Environmental Defense Fund, "Failure of Social Forestry in Karnataka"; Joyce, "Tree That Caused a Riot," p. 55; Kumar, "Eucalypts in Industrial and Social Plantations."

31. Environmental Defense Fund, "Failure of Social Forestry," p. 152; Lamb and Percy, "Indians Fight Eucalyptus Plantations"; Chandrashekhar, Krishnamurti, and Ramaswamy, "Social Forestry in Karnataka."

32. World Bank, *Forest Sector,* pp. 77–82; Environmental Defense Fund, *Statement* 1990, p. 6.

33. Saxena, "Marketing Constraints."

34. Saxena, *India's Eucalyptus Craze.*

35. Ravindranath, Gadgil, and Campbell, "Ecological Stabilization."

36. Cernea, "Sociological Framework."

37. Bhattacharya, *Social Forestry*, pp. 96–97, 100.

38. Patil, "Local Communities."

39. Saxena, *India's Eucalyptus Craze*.

40. Kumar, "Eucalypts in Industrial and Social Plantations," p. 28; Prasad and Ramaswamy, "Social Implications," p. 36.

41. Joshi, *Troup's The Silviculture of Indian Trees*, 5:169–72; Uniyal, "India-Environment"; Bose, "Shrinking Forests of India." Rajan, *Versatile Eucalyptus*, takes a very positive position about the benefits of the genus, concluding that "the advantages and uses are so many as to outweigh the limitations . . . enormously" (p. 201).

Chapter 8 / Environmental Issues

1. Shah, "Eucalyptus—Friend or Foe?" p. 195. Zobel, "Eucalyptus in the Forest Industry," regards environmental concerns in the minds of laymen as misplaced, but admits that these "bad boys of forestry" (p. 46) do take out croplands and may not be as prized over the long term.

2. Evans, *Plantation Forestry*, p. 23. The positive regard for plantations comes from the late 1960s and early 1970s.

3. Ball, "Development of Eucalyptus Plantations," p. 25.

4. Evans, *Plantation Forestry*, p. 30.

5. Ibid., pp. 13–15; Ball, "Development of Eucalyptus Plantations," p. 24.

6. Evans, *Plantation Forestry*, p. 21; Ball, "Development of Eucalyptus Plantations," p. 18; Foster, "Business and Environment."

7. A useful survey of the benefits and costs is provided by Evans, *Plantation Forestry*. Boland, "Australian Trees in Five Overseas Countries," describes the Cerrado as "one of the world's last great agricultural frontiers" (p. 25), from which Brazil already obtains 45 percent of its rice, 30 percent of beef, and 25 percent of coffee.

8. Evans, *Plantation Forestry*, pp. 340–43, gives a well-documented overview of the discussion about ecological diversity, stability, and vulnerability to pests. See Pryor, "Some Problems," for earlier remarks about growing eucalypts in low-latitude lowlands of Africa and Latin America, including comments on stem canker and harvester ants.

9. Goor, "Establishment, Management and Protection of Eucalypts," p. 36; Carne and Taylor, "Insect Pests"; Eldridge et al., *Eucalypt Domestication and Breeding*, p. 156.

10. Eldridge et al., *Eucalypt Domestication and Breeding*, p. 112.

11. Senanayake, "Analog Forestry."

12. Evans, *Plantation Forestry*, pp. 334–40; Duff et al., "Survey of Wildlife"; Poore and Fries, *Ecological Effects*, pp. 48–50; Ken Eldridge, personal communication, September 1998.

13. "The Koala," in Grzimek, ed., *Grzimek's Animal Life Encyclopedia*, 1:125;

Nowak, *Walker's Mammals of the World,* pp. 63–64. Bounds et al., "Observations," detail the food sources, nest sites, and foraging behaviors of a highly esteemed species in North Watson, Canberra, Australia.

14. Smith, "Utilization of Gum Trees," p. 155.

15. Ibid., p. 160.

16. Cited in Poore and Fries, *Ecological Effects,* p. 49.

17. Lyons, "Bird Use of Human-Disturbed Habitat," pp. 110–16.

18. Recher, "Eucalyptus Forests, Woodlands and Birds."

19. Poore and Fries, *Ecological Effects,* p. 48. See Kordell et al., "Eucalyptus in Portugal." FAO, *The Eucalypt Dilemma,* p. 18; Australian Centre for International Agricultural Research, *Eucalypts.*

20. Birla Institute of Scientific Research, *Social Forestry in India,* p. 52.

21. Poore and Fries, *Ecological Effects.* Evans, *Plantation Forestry,* refers to Poore and Fries, and accepts the tree's controversial status. He declares that "to exclude such an immensely useful genus will deny countless peoples across the tropics a valuable pole and fuel tree quite apart from industrial roles" (p. 339).

22. Poore and Fries, *Ecological Effects,* p. 54.

23. Ibid., p. 54.

24. Zobel, van Wyk, and Stahl, *Growing Exotic Forests,* p. 393.

25. Florence, "Cultural Problems."

26. I. R. Calder, "Water Use of Eucalypts—A Review," in Calder, Hall, and Adlard, eds., *Growth and Water Use of Forest Plantations,* p. 174; ACIAR, *Eucalypts*; Birla Institute of Scientific Research, *Social Forestry in India,* pp. 50–60.

27. Kordell et al., "Eucalyptus in Portugal," p. 10.

28. Davidson, "Ecological Aspects of *Eucalyptus* Plantations," pp. 43–48, 57–58; White, "Silviculture of *Eucalyptus* Plantings," pp. 78–79, citing a study by Ong (1993). Cremer, Cromer, and Florence, "Stand Establishment," p. 98, refer to "alleopathic" eucalypts, which "suppress their own progeny by chemical inhibitors leached from the trees' roots or litter."

29. Florence, "Cultural Problems," pp. 150–52.

30. Ibid., pp. 141, 143; Eldridge et al., *Eucalypt Domestication and Breeding,* p. 3.

Chapter 9 / A Global Solution: International Agencies Promote Eucalypts

1. *Encyclopedia of Associations: International Organizations 1992: Part One: Descriptive Listings,* p. 6060.

2. For background on Jacobs, see Meyer, *Foresters.*

3. E. Saouma, "Foreword," in Jacobs, *Eucalypts for Planting,* p. v.

4. Ibid. Joshi, *Troup's The Silviculture of Indian Trees,* 5:163, notes that Métro's book "rendered a signal service in the dissemination of information regarding the genus Eucalyptus and its various species."

5. S. J. Midgley, "Australian Tree Seed Centre"; Brown, "Global Use." For a discussion of the establishment of CSIRO in 1926, see Carron, *History of Forestry in Australia.*

6. Boland, "Eucalyptus Seed for Indian Plantations"; Vercoe, "Australian Trees on Tour."

7. ACIAR, *Improving and Sustaining Productivity.*

8. See Jiayu, "Genetic Improvement"; Rothschild, "Foreword," p. 5; De la Cruz, Lorilla, and Aggangan, "Ectomycorrhizal Tablets."

9. Streets, *Exotic Forest Trees,* "Foreword"; Troup, *Exotic Forest Trees.*

10. See FAO, *First World Eucalyptus Conference, Final Report,* pp. 19–27, 69, 119–27.

11. Ibid., pp. 91–97.

12. Ibid., p. 55.

13. Métro, "Round-up of Research Progress," pp. 26–27, 55–57 (conference recommendations).

14. A recommendation from the First World Eucalyptus Conference had asked the FAO directorate to set up a special working group on eucalypts for tropical Africa, in liaison with the Commission for Technical Co-operation in Africa South of the Sahara.

15. FAO, *Second World Eucalyptus Conference,* pp. 7–8, 12–13, 22–23, 31–33; Jacobs, *Genus Eucalyptus in World Forestry,* p. 15.

16. Examples are the Third World Consultation on Forest Tree Breeding, Canberra, 1978; IUFRO Symposium on Site and Productivity of Fast Growing Plantations, Pretoria, 1984; Nineteenth Meeting of the Canadian Tree Improvement Association, Toronto, 1985; Management of the Forests of Tropical America, Rio Pedras, Puerto Rico, 1986; Breeding Tropical Trees, IUFRO Conference, Pattaya, Thailand, 1989; Conference on Fast Growing Trees and Nitrogen Fixing Trees, Marburg, Germany, 1989; Regional Forest Agreements and the Public Interest, Canberra, Australia, 1998.

17. Eldridge et al., *Environmental Management,* p. iii; "Ectomycorrhizal Fungi," p. 2.

18. Zobel, Van Wyk, and Stahl, *Exotic Forests,* pp. 21–24, 409–11.

19. Eldridge et al., *Eucalypt Domestication and Breeding,* p. 111.

Conclusion

1. Dixit, "Development"; Boyd, "Bangkok to Ease Curbs"; Puntasen, "Political Economy of Eucalyptus."

2. Magagnini, "Leads Fight." A useful assessment of the significance of religion in respect to forest clearances is given by Sponsel and Natadecha, "Buddhism, Ecology and Forests in Thailand."

3. D. Price, "Cash Crop Fights Eucalyptus"; D. Price, "Phoenix Removes Pollution Risk"; Lohmann, "Commercial Tree Plantations," p. 13.

4. T. Gill, "Thailand-Environment."

5. Sargent, "Natural Forest," p. 20.

6. "Thai Government Studying Protesters' Complaints"; Mydans, "As Turmoil Builds"; Janchitjah, "Villagers Need Rice"; "Poor Year Seen."

7. Betts and Smith, "Utilization of California Eucalypts," pp. 24, 26; Martin, "Benefits of Hybridization."

8. Florence, *Ecology and Silviculture of Eucalypt Forests,* pp. 107–9.

9. Rodebaugh, "Demand Firm for Baby Eucalypts"; "Better Plants."

10. Evans, *Plantation Forestry,* p. 26. Jacobs, *Genus Eucalyptus in World Forestry,* p. 8, names fourteen species, including *E. globulus,* which became the more or less official tree.

11. Evans, *Plantation Forestry,* pp. 35–36; Jacobs, *Eucalypts for Planting,* pp. 76–77; Fortmann, "Great Planting Disasters," p. 49.

12. Poffenberger, "Resurgence," pp. 354–60.

13. Gadgil and Guha, *Ecology and Equity,* pp. 162–67; Gadgil and Guha, *This Fissured Land.*

14. Panayotou and Ashton, *Not by Timber Alone,* pp. 171–81.

15. Evans, *Plantation Forestry,* p. 32; Kanowski, "Australian Forestry," p. 2.

References

ABECEL. *Statistical Report.* Rio de Janeiro: ABECEL, 1992.

Algar, W. H. *Forestry and Forest Products.* Parkville: Australian Academy of Technological Sciences, 1988.

Aracruz Celulose S.A. *Annual Report* (by year). Rio de Janeiro: Aracruz, 1980, etc.

Arbenz, J. H. *The Eucalyptus in Texas.* Texas Department of Agriculture Bulletin 8. 1911.

Argentine Republic, Ministerio de Económica. *Argentina: A Potential Producer of Paper and Pulp.* Buenos Aires: Ministerio de Económica, 1974.

"Assessment of Alternative Pulping and Bleaching Technologies." *World Paper* 220, no. 9 (October 1995): 24.

Aubreville, A. "Quelques problèmes forestiers du Brésil." *Bois et forêts tropicales* 6 (1948): 102–17.

Australian Bureau of Agricultural and Resource Economics. *Quarterly Forest Products Statistics.* Canberra: ABARE, 1994.

Australian Centre for International Agriculture Research. *Eucalypts: Curse or Cure?* Canberra: ACIAR, 1992.

———. *Improving and Sustaining Productivity of Eucalypts in SouthEast Asia.* Canberra: CSIRO, 1995.

———. "Wood Production from Australian Trees in China." *Research Notes* 5 (1990): 1–4.

Australian Dictionary of Biography. Vols. 4–5. Melbourne: Melbourne University Press, 1972.

Bail, Murray. *Eucalyptus: A Novel.* New York: Farrar, Straus, Giroux, 1998.

Baker, Richard T., and Henry G. Smith. *A Research on the Eucalypts and Their Essential Oils.* Sydney: Government of New South Wales, 1920.

Ball, J. B. "Development of Eucalyptus Plantations—An Overview." In *Proceedings of the Regional Expert Consultation on Eucalyptus, October 1993* (Bangkok: FAO, 1995), 15–27.

Banham, P. W., K. Orme, and S. L. Russell. "Pulpwood Qualities Required for the Cold Soda Pulping Process." In *Eucalypt Plantations: Improving Fibre Yield and Quality,* ed. B. M. Potts et al. Proceedings of the CRC-IUFRO Conference, Hobart, February 1995. Hobart: Co-Operative Research Center for Hardwood Forestry, 1995. 1–4.

References

Beaglehole, J. C., ed. *The* Endeavour *Journal of Joseph Banks (1768–1771).* 2 vols. Melbourne: Angus and Robertson, 1962.

Bentham, George, and Ferdinand von Mueller. *Flora Australiensis.* London: Reeves, 1866.

"Better Plants." *USA Today,* 23 June 1998.

Betts, H. S., and C. Stowell Smith. "Utilization of California Eucalypts." USDA, Forest Service, Circular 179. Washington, D.C.: Government Printing Office, 1910.

Bhattacharya, Pranab Kumar. *Social Forestry: A Step toward Environmental Change.* New Delhi: Khama, 1990.

Birla Institute of Scientific Research. *Social Forestry in India: Problems and Prospects.* New Delhi: Radiant, 1984.

Blair, Pat. "Deadly Buzz That Warns of Disease." *(London) Times,* 14 November 1994.

"Blue Gum." *Kew Bulletin* (1903): 1–10.

Boland, Douglas J. "Australian Trees in Five Overseas Countries." Churchill Fellowship Report. Canberra: CSIRO, 1984.

———. "Brief History of the Eucalyptus Oil Industry." In *Eucalyptus Leaf Oils,* ed. D. J. Boland et al. (Melbourne: Inkata Press, 1991), 3–10.

———. "Eucalyptus Seed for Indian Plantations from Better Australian Natural Seed Sources." *Indian Forester* 107 (1981): 125–34.

———. "Galloping Gums." *Forestry Log* 15 (1982): 27–33.

Boland, Douglas J., et al. *Eucalyptus Seed.* Canberra: CSIRO, 1980.

Boland, Douglas J., et al., eds. *Forest Trees of Australia.* Melbourne: Nelson, 1984.

Booker, M. I. H., and D. A. Kleinig. *Field Guide to Eucalypts of South-Eastern Australia.* Melbourne: Inkata Press, 1993.

Bose, Kunal. "The Shrinking Forests of India." *Financial Times,* 6 September 1995.

Bosisto, Joseph. *The Indigenous Vegetation of Australia.* London: Clowes, 1886.

———. "Is the Eucalyptus a Fever-Destroying Tree?" *Proceedings of the Royal Society of Victoria* 12 (1875): 10–23.

Bounds, Jenny, et al. "Observations of a Breeding Colony of Four Pairs of Regent Honeyeaters." *Canberra Bird Notes* 21 (1996): 41–55.

Boyd, A. "Bangkok to Ease Curbs on Commercial Timber Farming." *Business Times,* 29 June 1993.

[Brazil.] *Irish Times,* 7 May 1993.

[Brazil's Pulp Output.] *Paper and Pulp International* 40 (1998): 41.

Brockway, Lucille. *Science and Colonial Expansion.* New York: Academic Press, 1979.

Brooker, M. I. H., and D. A. Kleinig. *Field Guide to Eucalypts.* 3 vols. Melbourne: Inkata Press, 1983.

Brown, A. G. "Global Use of Australian Genetic Resources." In *Australasian Forestry and the Global Environment,* ed. R. N. Thwaites and B. J. Schaumberg (Alexandra Headland: Institute of Foresters of Australia, 1993), 179–86.

Burgess, I. P. "The Natural Occurrence of *Eucalyptus grandis,* Its Distribution Patterns in Natural Forests." *Silvicultura* 31 (1983): 397–99.

References

Calder, Ian R., R. L. Hall, and P. G. Adlard, eds. *Growth and Water Use of Forest Planta-tions.* Chichester: Wiley, 1992.

Carne, P. B., and K. L. Taylor. "Insect Pests." In *Eucalypts for Wood Production,* ed. Hillis and Brown, 155–68.

Carrere, Ricardo. "Pulping the South: Brazil's Pulp and Paper Plantations." *Ecologist* 26 (1996): 206–15.

Carrere, Ricardo, and Larry Lohmann. *Pulping the South.* London: Zed Books, 1996.

Carron, L. T. *A History of Forestry in Australia.* Canberra: Australia National University, 1985.

Cat Spring Agricultural Society. *A Century of Agricultural Progress.* Austin County, Texas: Cat Spring Agricultural Society, 1956.

Cavalcanti, Cristina. "Brazil's New Forests Bring Profit and Pain." *People and Planet* 5 (1996): 14–16.

Cernea, Michael M. "A Sociological Framework." In *Managing the World's Forests,* ed. Narendra S. Sharma (Dubuque: Kendall Hunt, 1992), 314–17.

Chandrashekhar D. M., B. V. Krishnamurti, and S. R. Ramaswamy. "Social Forestry in Karnataka: An Impact Analysis." *Economic and Political Weekly* 22 (June 1987): 935–41.

Chisholm, Alec H. *Ferdinand von Mueller.* Melbourne: Oxford University Press, 1962.

Chown, T. C. "Raising Fever Gum Trees." *Garden,* 22 May 1874, 434.

Churchill, D. M., T. B. Muir, and D. M. Sinkora. "The Published Works of Ferdinand J. H. Mueller (1825–1896)." *Muelleria* 4 (1978): 1–120.

Clapp, Roger A. "The Unnatural History of the Monterey Pine." *Geographical Review* 85 (1995): 1–19.

Cody, Harold M. "Aracruz Continues Leadership." *Pulp and Paper,* 1 October 1998, 55.
———. "Brazil's No. 3 Eucalyptus Producer." *Pulp and Paper,* 1 November 1998, 67.

Cohn, Helen M. "Ferdinand Mueller, Government Botanist: The Role of William Hooker in His Appointment." *Muelleria* 7 (1989): 99–102.

"Colcura: Leader in Eucalyptus Cultivation." *Chilean Forestry News* 11 (1988): 5–6.

Cook, Thomas C. "The Distribution of Certain Species of Exotic Eucalyptuses." M.A. thesis, University of Texas, Austin, 1959.

Cooper, Ellwood. *Forest Culture and Eucalyptus Trees.* San Francisco: Cubery, 1876.

Couto, L., and D. R. Betters. "Short-Rotation Eucalypt Plantations in Brazil." World Wide Web site: http://www.esd.ornl.gov/BFDP/BFDP. 23 September 1996.

Cremer, K. W., R. N. Cromer, and R. G. Florence. "Stand Establishment." In *Eucalypts for Wood Production,* ed. Hillis and Brown, 81–135.

"Crisis: Eucalypts, a Damaged Forest." Berkeley: University of California, Department of Geography, 1973. Mimeo.

Cromer, Robin N. "The Role of Eucalypt Plantations in Timber Supply and Forest Conservation in Australia." In *Second IUFRO, XIX World Forestry Congress* (n.p.: Science in Forestry, 1991), 299–310.

Dargavel, John, and Noel Semple, eds. *Prospects for Australian Forest Plantations.* Canberra: CRES, 1990.

Davidson, John. "Ecological Aspects of *Eucalyptus* Plantations." In *Proceedings of the Regional Expert Consultation on Eucalypts,* 3 vols. (Bangkok: FAO, 1995), 1:35–72.

Dean, G. H. "Objectives for Wood Fibre Quality and Uniformity." In *Eucalypt Plantations: Improving Fibre Feed and Quality,* ed. B. M. Potts et al. (Hobart: Co-Operative Research Center for Hardwood Forestry, 1995), 5–9.

De la Cruz, R. E., E. B. Lorilla, and N. S. Aggangan. "Ectomycorrhizal Tablets for *Eucalyptus* Species." In *Fast Growing Trees and Nitrogen Fixing Trees,* ed. D. Werner and P. Müller (Stuttgart: Fischer Verlag, 1990), 371.

De Miranda Bastos, A. "Importance of Eucalypts in Brazil." In FAO, *Second World Eucalyptus Conference,* 2:1068–69.

De Vilmorin, Henry. "Flowers on the French Riviera." *Journal of the Royal Horticultural Society* 15 (1894): 80–104.

Dixit, Kunda. "Development: Greens Rap IMF–World Bank's 'Destructive' Vision." Inter Press Service, 10 October 1991.

D'Ombrain, A. W. *A Gallery of Gum Trees.* Sydney: Australian Medical Publishing, 1938.

Doran, John C. "Commercial Sources, Uses, Formation, and Biology." In *Eucalyptus Leaf Oils,* ed. D. J. Boland et al. (Melbourne: Inkata Press, 1991), 11–25.

Doran, John C., and John W. Turnbull, eds. *Australian Trees and Shrubs: Species for Land Rehabilitation and Farm Planting in the Tropics.* Canberra: CSIRO, 1998.

"Dr. Navarro de Andrade's Acceptance." *Journal of Heredity* 32 (1941): 213–14.

Drury, Herbert. *The Useful Plants of India.* 2nd ed. London: Allen, 1873.

Dücker, Sophie C. "Australian Phycology: The German Influence." In *People and Plants in Australia,* ed. D. J. and S. G. M. Carr (Sydney: Academic Press, 1981), 116–38.

Duff, A. B., et al. "A Survey of Wildlife in and around a Commercial Tree Plantation in Sabah." *Malaysian Forester* 47 (1984): 197–213.

Duval, Marguerite. *The King's Garden.* Charlottesville: University of Virginia Press, 1982.

Dwivedi, A. P. *Forestry in India.* Dehra Dun: Kishore, 1980.

"Ectomycorrhizal Fungi for Eucalypt Plantations in China." *ACIAR Newsletter* 21 (1996): 2.

Eldridge, Ken G. *Eucalyptus Camaldulensis.* Tropical Forestry Papers 8. Oxford: Commonwealth Forestry Institute, 1975.

Eldridge, Ken, et al. *Eucalypt Domestication and Breeding.* Oxford: Clarendon Press, 1994.

Eldridge, Ken, et al., eds. *Environmental Management: The Role of Eucalypts and Other Fast Growing Species.* Collingwood: CSIRO, 1996.

Elwes, Henry J., and Augustine Henry. *Trees of Great Britain and Ireland,* vol. 6. Edinburgh: Privately published, 1906–13.

Encyclopedia of Associations: International Organizations 1992: Part One: Descriptive Listings. Detroit: Gale, 1992.

Environmental Defense Fund. "The Failure of Social Forestry in Karnataka." *Ecologist* 17 (1987): 151–54.

———. Statement of Lori Udall before the Subcommittee on International Economic Policy, Trade, Oceans, and Environment, Committee of Foreign Relations, U.S. Senate, 18 July 1990. Washington, D.C.: EDF, 1990.

"Eucalypts in Scotland." *Transactions and Proceedings of the Botanical Society of Edinburgh* 60 (1896): 515–31.

"Eucalyptus for English Gardens." *Journal of the Royal Horticultural Society* 42 (1916): 192.

"Eucalyptus Timber for Street Paving." *Kew Bulletin* (1897): 219–21.

Evans, Julian. *Plantation Forestry in the Tropics.* 2nd ed. Oxford: Clarendon Press, 1992.

FAO. *The Eucalypt Dilemma.* Rome: FAO, 1988.

———. *First World Eucalyptus Conference, Final Report.* Rome: FAO, 1956.

———. *Pulping and Paper-Making Properties of Fast-Growing Plantation Wood Species.* Vol. 1. FAO Forestry Paper 19/1. Rome: FAO, Forest Industries Division, Forestry Department, 1980.

———. *Second World Eucalyptus Conference: Report and Documents.* 2 vols. São Paulo: FAO, 1961.

Fernow, B. E. *Some Foreign Trees for the Southern States.* Washington, D.C.: Government Printing Office, 1910.

Flannery, Timothy F. *The Future Eaters: An Ecological History of the Australasian Lands and People.* Chatswood, NSW: Reed, 1994.

Fletcher, Harold R. *The Story of the Royal Horticultural Society, 1804–1968.* London: Oxford University Press, 1969.

Flinta, C. M. "The Value of Eucalypts: Latin America." In FAO, *First World Eucalyptus Conference* (Rome: FAO, 1956), 76–83.

Florence, Ross G. "Cultural Problems of Eucalyptus as Exotics." *Commonwealth Forestry Review* 65 (1986): 141–63.

———. *Ecology and Silviculture of Eucalypt Forests.* Collingwood: CSIRO, 1996.

Fortmann, Louise. "Great Planting Disasters: Pitfalls in Technical Assistance in Forestry." *Agriculture and Human Values* 5 (1988): 49–60.

Foster, Angus. "Business and the Environment: Reality Beckons." *Financial Times,* 21 June 1995.

Fung, P. Y. H. *Eucalypts in Brazil for Charcoal.* Canberra: CSIRO, 1988.

Gadgil, M., and R. Guha. *History and Equity: The Uses and Abuses of Nature in Contemporary India.* London: Routledge, 1995.

———. *This Fissured Land: An Ecological History of India.* New Delhi: Oxford University Press, 1992.

Garden, The (England). 12 October 1872; 21 March and 2 May 1874; 19 February, 27 May, 8 July 1876; 29 September 1877; 2 and 16 November 1878; August 1880; 6 May 1882; 27 April 1889.

References

Gardener's Chronicle (England). 9 December 1843; 30 March 1844; 3 May and 21 June 1851; 7 March 1874.

Gardener's Magazine (England) 6 (1830): 502; 11 (1835): 151, 570; 13 (1837): 23, 463; 14 (1838): 511.

Gatti, Daniel. "Environment—Latam." Inter Press Service, 7 October 1998.

Ghate, Rucha S. *Forest Policy and Tribal Development.* New Delhi: Concept, 1992.

Gildas, Frère. "L'Eucalyptus dans la campagne romaine." *Bulletin de la Société d'Acclimatation* (1875): 180–85.

Gill, Teena. "Thailand-Environment: Log Ban Fails." Inter Press Service, 5 February 1996.

Golfari, Lamberto, Roberto L. Caser, and Vicente P. G. Moura. *Zoneamento ecológico equemático par reflorestamento no Brasil.* Belo Horizonte: Centro de Pesquisa Florestal, 1978.

Goor, A. Y. "Establishment, Management and Protection of Eucalypts." In FAO, *First World Eucalyptus Conference* (Rome: FAO, 1956), 29–37.

Griffith, Victoria. "Reversal of Fortune." *Latin Finance* 61 (1994): 65.

Grimwade, Russell. *An Anthology of Eucalypts.* Sydney: Angus and Robertson, 1920.

Groenendaal, Gayle M. "California's First Fuel Crisis and Eucalyptus Plantings." M.A. thesis, California State University, Northridge, 1985.

Grzimek, Bernhard, ed. *Grzimek's Animal Life Encyclopedia.* New York: Van Nostrand, 1972.

Hall, Cuthbert. *On Eucalyptus Oils.* Paramatta: Little, 1904.

Hall, Norman. *Botanists of the Eucalypts.* Melbourne: CSIRO, 1978.

Harcharik, David. "Foreword." In *Proceedings of the Regional Expert Consultation on Eucalyptus,* 3 vols. (Bangkok: FAO, 1995), 1:iii.

Heysen's Gum Trees. Sydney: Morning Herald, 1937.

Higgins, H. G. "Pulp and Paper." In *Eucalypts for Wood Production,* ed. Hillis and Brown, 290–316.

Higgs, Richard. "Going for Growth, Grupo Santa Fe." *World Paper* 220 (September 1995): 45–46.

Hillis, W. E. "Chemicals." in *Eucalypts for Wood Production,* ed. Hillis and Brown, 357–60.

———. "Fast Growing Eucalypts and Some of Their Characteristics." In *Fast Growing Trees and Nitrogen Fixing Trees,* ed. D. Werner and P. Müller (Stuttgart: Fischer Verlag, 1990), 184–93.

Hillis, W. E., and A. G. Brown, eds. *Eucalypts for Wood Production.* Sydney: Academic Press, 1984.

Holusha, John. "Pulp Mills Turn Over a New Leaf." *New York Times,* 9 March 1996.

Illman, John. "Happiness Is Needle-Shaped." *Guardian,* 21 January 1994.

Inchbald, Peter. "*Eucalyptus Globulus* on the Riviera." *Garden,* 6 March 1875, 202.

Ingham, Norman D. *Eucalyptus in California.* University of California, College of Agriculture, Agricultural Experiment Station, Bulletin 196. Sacramento: Shannon, 1908.

Jacobs, Maxwell R. "Eucalyptus as an Exotic." In *Second World Eucalyptus Conference* (Canberra: Forestry and Timber Bureau, 1961), 5.

———. *Eucalypts for Planting.* Rome: FAO, 1981.

———. *Forestry Development and Research, Brazil.* FAO Technical Report 1, FO:SF/BRA45. Rome: FAO, 1972.

———. *The Genus Eucalyptus in World Forestry.* Seattle: University of Washington Press, 1970.

Janchitjah, S. "Villagers Need Rice Not Eucalyptus." *Bangkok Post,* 30 November 1997.

"Jarrah Timber." *Kew Bulletin* (1890): 188–89.

Jean, G. H. "Objectives for Wood Fibre Quality and Uniformity." In *Eucalypts Plantations: Improving Fibre Yield,* ed. B. M. Potts et al. (Hobart: Co-op Research for Hardwood Forestry, 1995), 5–9.

Jiayu, B. "Genetic Improvement of Plantation Eucalyptus Tree Species in China." In *Australian Tree Special Research in China,* ed. A. G. Brown (Canberra: ACIAR, 1994), 32–49.

Joshi, H. B. *Troup's The Silviculture of Indian Trees.* Vol. 5. Delhi: Government of India Press, 1984.

Joyce, Christopher. "The Tree That Caused a Riot." *New Scientist,* 18 February 1988, 54–59.

Kanowski, Peter. "Australian Forestry in Transition." Paper presented at Australian Forest Growers Conference, Mt. Gambier, 1996.

Keast, A., et al. *Birds of Eucalyptus Forests and Woodlands.* Chipping Norton, NSW: RAOU, 1985.

Kelly, Stan. *Eucalypts.* 2 vols. New York: Van Nostrand, 1983.

King, James, and Stanly L. Krugman. *Tests of 36 Eucalyptus Species in Northern California.* Research Paper PSW-152. Berkeley: Pacific Southwest Forest and Range Experiment Station, 1980.

Kingzett, C. T. *Nature's Hygiene.* London: Bailliere, Tindale and Cox, 1907.

Kinney, Abbot. *Eucalyptus.* Los Angeles: Baumgardt, 1895.

Kordell, Lars, et al. "Eucalyptus in Portugal." *Ambio* 15 (1986): 6–13.

Kumar, P. J. Dilip. "Eucalypts in Industrial and Social Plantations in Karnataka." In *Growth and Water Use of Forest Plantations,* ed. Calder, Hall, and Adlard, 16–29.

Kumar, Vinod. "Eucalyptus in the Forestry Scene of India." In *Intensive Forestry: The Role of Eucalypts,* ed. A. P. G. Schonau, 2 vols. (Pretoria: South Africa Institute of Forestry, 1991), 2:1106.

Kynaston, Edward. "Exploration as Escape: Baron Sir Ferdinand von Müller." In *From Berlin to Burdekin,* ed. David Walker and Jürgen Tampke (Kensington: NSW University Press, 1991), 3–21.

———. *A Man on the Edge.* Ringwood: Allen Lane, 1981.

"Ladrões Roubam Eucalipto." *Municípios* (Bahia), 13 September 1994.

Lamb, Harriet, and Steve Percy. "Indians Fight Eucalyptus Plantations on Commons." *New Scientist,* 16 July 1987, 31.

Landsborough, D. "Eucalypts in Scotland." *Transactions of the Botanical Society of Edinburgh* (1896): 515–31.

Lavery, P. B., and D. J. Mead. "*Pinus radiata:* A Narrow Endemic from North America Takes on the World." In *Ecology and Biogeography of* Pinus, ed. D. M. Richardson (Cambridge: Cambridge University Press, 1998), 432–49.

Lawless, Julia. *The Encyclopedia of Essential Oils.* Shaftsbury, Dorset: Element, 1992.

Lewington, Anna. *Plants for People.* New York: Oxford University Press, 1990.

Little, Elbert L. "Fifty Trees from Foreign Lands." In *USDA Yearbook for 1949* (Washington, D.C.: Government Printing Office, 1949), 815–32.

Livernash, Robert. "India." In *World Resources, 1994–95* (New York: Oxford University Press, 1994), 83–106.

Lockwood, Samuel. "The Eucalyptus in the Future." *Popular Science Monthly* 12 (1878): 662–71.

Lohmann, Larry. "Commercial Tree Plantations in Thailand." *Ecologist* 20 (1990): 9–17.

Loudon, J. C. *Arboretum et Fruticetum Britannicum.* London: Longman, 1838.

———. *Hortus Britannicus.* 2nd ed. London: Longman, 1832.

Lucas, A. M. "Baron von Mueller: Protégé Turned Patron." In *Australian Science in the Making,* ed. R. W. Home (Cambridge: Cambridge University Press, 1988), 133–52.

Lyons, Jane T. "Bird Use of Human-Disturbed Habitat: The Implications for Management and Policy." Ph.D. dissertation, University of Texas, Austin, 1996.

Magagnini, Stephen. "Leads Fight for Thai Forests." *Sacramento Bee,* 29 November 1992.

Maiden, J. H. *A Bibliography of Australian Economic Botany.* Sydney: Potter, 1892.

———. *A Critical Review of the Genus Eucalyptus.* 2 vols. Sydney: Government of New South Wales, 1909.

———. *Sir Joseph Banks.* Sydney: Gullick, 1909.

Malhotra, P. P., and K. K. Sharma. "Need for Defining Wasteland for Afforestation Purposes." In *Wasteland for Development and Fuelwood and Fodder Production* (Delhi: F.R.I. Press, 1988), 3–9.

Marcus Wallenberg Foundation. *The New Eucalypt Forest: Proceedings of the Marcus Wallenberg Foundation Symposia.* Lectures given by 1984 Marcus Wallenberg prize winners in Falun, Sweden, 14 September 1984. Falun: Marcus Wallenberg Foundation, 1984.

Maroske, Sara, and H. M. Cohn. "'Such Ingenious Birds': Ferdinand Mueller and William Swainson in Victoria." *Muelleria* 7 (1992): 529–53.

Martin, B. "The Benefits of Hybridization." In *Proceedings of IUFRO Conference,* Pattaya, Thailand, November 1988, ed. G. L. Gibson, A. R. Griffin, and A. C. Matheson (Oxford: Oxford Forestry Institute, 1989), 79–92.

Mathur, R. B. "Eucalyptus." In Forest Research Institute, *Proceedings of First Forestry Conference, December 6–10, 1973* (Dehra Dun: F.R.I., 1979), 1:21.

Mathur, R. S., S. R. Sagar, and M. Y. Ansari. "Economics of *Eucalyptus* Plantations— with Special Reference to Uttar Pradesh." *Indian Forester* 110 (1984): 97–109.

McClatchie, Alfred J. *Eucalypts Cultivated in the United States.* USDA, Bureau of Forestry, Bulletin 35. Washington, D.C.: Government Printing Office, 1902.

McIlroy, J. C. "Grazing Animals." In *Eucalypts for Wood Production,* ed. Hillis and Brown, 139–42.

Metcalf, Woodbridge. "Eucalyptus Trees in the United States." In *First World Eucalyptus Conference* (Rome: FAO, 1956), 91–96.

———. *Growth of Eucalyptus in California Plantations.* California Agricultural Experiment Station Bulletin 380. Berkeley: University of California Press, 1924.

Métro, André. *Eucalypts for Planting.* Rome: FAO, 1955.

———. "Round-up of Research Progress in the Past Five Years and Future Needs." In *First World Eucalyptus Conference* (Rome: FAO, 1956), 19–27.

Meyer, Athol. *The Foresters.* Hobart: Institute of Foresters of Australia, 1985.

Midgley, S. J. "The Australian Tree Seed Centre." In *International Forestry Conference for the Australian Bicentenary 1988,* 5 vols. (Albury: AFDI, 1988), 5:1–9.

Midgley, S. J., et al. "Exotic Plant Species in Vietnam's Economy." Seminar paper on environment and development in Vietnam, 6–7 December 1996, Australian National University, Canberra.

Montalbano, William. "'Fascist' Trees Cause Furor in Iberia." *Toronto Star,* 23 February 1991.

Moore, James. "Green Gold: The Riches of Baron Ferdinand von Mueller." Paper presented at Royal Botanic Gardens, 1996 Commemorative Conferences, Melbourne.

Moulds, Frank R. "Eucalypts and Their Use in Semi-Tropical Plantings." *Tropical Woods* 91 (1947): 1–16.

Moyal, Ann. *A Bright and Savage Land.* Ringwood: Penguin, 1986.

Mydans, Seth. "As Turmoil Builds, Thai Leader Shuffles Cabinet." *New York Times,* 25 October 1997.

Naudin, Charles. *Description et emploi des Eucalyptus.* Antibes: Marchand, 1891.

———. "Mémoire sur les Eucalyptus." *Annales des sciences naturelles—Botanique* 16 (1883): 337–430.

Navarro de Andrade, Edmundo. *O Eucalipto.* São Paolo: Chacaras e Quintais, 1939.

———. "The Eucalyptus in Brazil." *Journal of Heredity* 32 (1941): 215–20.

Nowak, Ronald M. *Walker's Mammals of the World.* 5th ed. Baltimore: Johns Hopkins University Press, 1991.

O'Connor, J. E. "The Cultivation of 'Eucalyptus Globulus' and other Australian Gums in India." *Indian Forester* 2 (1872): 120–36.

"On Ornamental Trees and Shrubs in the Garden of Castewellan." *Journal of the Royal Horticultural Society* 27 (1902–3): 407–27.

Panayotou, T., and P. S. Ashton. *Not by Timber Alone.* Washington, D.C.: Island Press, 1992.

Patanapongsa, N. "Private Forestry Developments in Thailand." *Commonwealth Forestry Review* 69 (1990): 63–68.

References

Patil, V. "Local Communities and *Eucalyptus*: An Experience in India." In FAO, *Proceedings of the Regional Expert Consultation on Eucalyptus,* 3 vols. (Bangkok: FAO, 1995), 1:129–36.

Penfold, A. R., and J. L. Willis. *The Eucalypts.* New York: Interscience, 1961.

Pepper, Edward. "Eucalyptus in Algeria and Tunisia from an Hygienic and Climatological Point of View." *Proceedings of the American Philosophical Society* 35 (1896): 39–56.

Phillips, F. H. "Manufacture of Pulp and Paper." In "China-Australian Afforestation Project at Dongmen State Forest Farm," 1–16. Technical Communication 40. Fourth Technical Exchange Seminar, 20–24 October 1989, Guangxi Zhuang Autonomous Region, PRC. Brisbane: Queensland Department of Forests, 1989. Mimeo.

Picornell, Pedro M. "Why Do Developing Countries Go into the Pulp and Paper Industry?" In *Establishing Pulp and Paper Mills,* 3–5. FAO, Forestry Paper 45. Rome: FAO, 1983.

Planchon, J. E. Eucalyptus Globulus *from a Botanic, Economic and Medical Point of View.* USDA Report 9. Washington, D.C.: Government Printing Office, 1875.

Poffenberger, Mark. "The Resurgence of Community Forest Management." In *Nature, Culture, Imperialism: Essays on the Environmental History of South Asia,* ed. David Arnold and Ramachandra Guha (Delhi: Oxford University Press, 1995), 336–69.

Poffenberger, Mark, and Betsy McGean, eds. *Village Voices, Forest Choices.* New Delhi: Oxford University Press, 1996.

Poore, Duncan, P. G. Adlard, and M. Arnold. "Environmental Implications of Eucalypts." In *The International Forestry Conference for the Australian Bicentenary 1988,* 5 vols. (Albury: Forest Development Institute, 1988), 1:1–9.

Poore, M. E. D., and C. Fries. *The Ecological Effects of Eucalyptus.* FAO Forestry Paper 59. Rome: FAO, 1985.

"Poor Year Seen." *Business Day* (Thailand), 5 February 1999.

Prasad, G. N. N., and S. R. Ramaswamy. "Social Implications of Eucalyptus Propagation." In *Growth and Water Use of Forest Plantations,* ed. Calder, Hall, and Adlard, 33–36.

Pratt, Merritt B. *The Use of Lumber on California Farms.* California Agricultural Experiment Station Bulletin 299. Berkeley: University of California Press, 1918.

Price, D. "Cash Crop Fights Eucalypts." *Benn Publications* 219 (1994): 32.

———. "Phoenix Removes Pollution Risk." *Benn Publications* 219 (1994): 33.

Price, Shirley. *The Aromatherapy Workbook.* London: Thorsons, 1993.

"Procès-Verbaux." *Bulletin de la Société d'Acclimatation* (1872): 206–8.

"Producers Foresee Increase." *Gazeta Mercantil,* 21 January 1997.

Pryor, L. D. *The Biology of Eucalypts.* London: Arnold, 1976.

———. "Some Problems of Eucalypt Plantations Overseas." *Forestry Log* 11 (1978): 36–37.

References

Pryor, L. D., W. G. Chandler, and B. Clarke. *The Establishment of Eucalyptus Plantations for Pulpwood Production in the Coffs Harbour Region of New South Wales.* Bulletin 1. Coffs Harbour: APM Forests Property, 1968.

Pryor, L. D., and L. A. S. Johnson. *A Classification of Eucalypts.* Canberra: Australian National University, 1976.

———. "Eucalyptus: The Universal Australian." In *Ecological Biogeography of Australia,* ed. Allen Keast (The Hague: Junk, 1981), 499–536.

[Pulp.] *Money,* October 1995.

Puntasen, Apichai. "Political Economy of Eucalyptus: Business, Bureaucracy, and the Thai Government." *Journal of Contemporary Asia* 22 (1992): 187–206.

Rajan, B. K. C. *Versatile Eucalyptus.* Bangalore: Diana, 1987.

Ramel, P. "*L'Eucalyptus Globulus.*" *Bulletin de la Société Impériale Zoologique d'Acclimatation* 9 (1862): 787.

Ravindranath, N. H., Madhar Gadgil, and Jeff Campbell. "Ecological Stabilization and Community Needs," in *Village Voices, Forest Choices,* ed. Poffenberger and McGean, 287–324.

Recher, Harry. "Eucalyptus Forests, Woodlands and Birds: An Introduction." In A. Keast et al., *Birds of Eucalypt Forests and Woodlands* (Chipping Norton, NSW: Royal Australasian Ornithologists Union, 1985), 1–10.

Redhead, J. F., and N. C. E. Anandarajah. "The Planting of Eucalyptus on Tea Estates." In UNDP/FAO, *Agriculture Diversification Project* (Peradeniya: Suhauk, 1976), 1–15.

Robinson, William. *The Subtropical Garden.* London: Murray, 1871.

Rocha, Jan. "Brazilian Court Halts 'Green' Firm's Forestry." *Guardian,* 26 November 1993.

Rodebaugh, Dale. "Demand Firm for Baby Eucalypts." *Chicago Tribune,* 14 January 1996.

Rothschild, G. H. L. "Foreword." In *Australian Tree Species Research in China,* ed. A. G. Brown (Canberra: Australian Centre for International Agricultural Research, 1994), 5.

Santos, Robert L. *The Eucalyptus of California: Seeds of Good or Seeds of Evil?* Denair: Alley-Cass, 1997.

Sargent, Caroline. "Natural Forest as Plantation?" In *Plantation Politics,* ed. C. Sargent and Stephen Bass (London: Earthscan, 1992), 20.

Saxena, N. C. *India's Eucalyptus Craze: The God That Failed.* New Delhi: Sage, 1994.

———. "Marketing Constraints for Eucalyptus from Farm Lands in India." *Agroforestry Systems* 13 (1991): 73–85.

Scarascia-Mugnozza, G., et al. "Freezing Mechanisms, Acclimation Processes and Cold Injury in *Eucalyptus* Species." *Forest Ecology and Management* 29 (1989): 81–94.

Schreuder, Gerard F., A. A. A. De Barros, and Denny A. Hill. "The Global Supply and Cost Price Structure of *Eucalyptus.*" In *Global Resources and Markets,* ed. Donald F. Root (Seattle: University of Washington Press, 1991), 81–92.

References

Sellers, C. H. *Eucalyptus: Its History, Growth and Utilization.* Sacramento: Johnston, 1910.

Senanayake, F. Ranil. "Analog Forestry as a Conservation Tool." *Tiger Paper* 14 (1987): 25–29.

Shah, S. A. "Eucalyptus—Friend or Foe?" *International Tree Crops Journal* 3 (1985): 191–95.

Sharma, Narendra, et al. "World Forests in Perspective." In *Managing the World's Forests,* ed. Narendra P. Sharma (Dubuque: Kendall Hunt, 1992).

Sharman, Peter. "High-yielding Pulp." *Canadian Papermaker* 48 (1995): 37.

Shiva, Vandana, and J. Bandyopadhyay. "Eucalyptus—A Disastrous Tree for India." *Ecologist* 13 (1983): 184–87.

Singh, Rudra P. "Pulping Horizons, Pulping Technology." *World Paper* 219 (June 1994): 31.

Smith, David. *Saving a Continent: Towards a Sustainable Future.* Sydney: UNSW Press, 1994.

Smith, K. D. "The Utilization of Gum Trees by Birds in Africa." *Ibis* 116 (1974): 155–64.

Sponsel, L. E., and P. Natadecha. "Buddhism, Ecology and Forests in Thailand." In *Changing Tropical Forests,* ed. F. Dargavel et al. (Canberra: CRES, 1988), 305–26.

Stafleu, Frans A., and Richard S. Cowan. *Taxonomic Literature.* 2nd ed. Vol. 2. The Hague: Junk, 1979.

Stefan, Virginia. "Market Pulp Producers Ride Wave." *Pulp and Paper* 69 (1995): 87–92.

Stier, Jeffrey C. "The World Eucapulp Industry." *Northern Journal of Applied Forestry* 7 (1990): 158–63.

Stoeckeler, Joseph H., and Ross A. Williams. "Windbreaks and Shelterbelts." In *USDA Yearbook for 1949* (Washington, D.C.: Government Printing Office, 1949), 191–99.

Streets, R. J. *Exotic Forest Trees in the British Commonwealth.* Oxford: Clarendon Press, 1962.

Sunder, S. "India." In FAO, *Forestry Policies in Selected Countries in Asia and the Pacific* (Rome: FAO, 1993), 23–35.

"Sustainable Development." *Institutional Investor,* 29 March 1994.

Sutton, Peter, John Pearson, and Hugh O'Brian. "World Paper Production Hits Record Level." *Pulp and Paper* 63 (August 1989): 58–76.

Swann, Charles E. "Market Pulp Powerhouse: Brazil." *American Papermaker* 57 (1994): 22–32.

Teicher, Johannes. "Manufacture of Newsprint with Eucalyptus Pulp." In FAO, *Second World Eucalyptus Conference: Report and Documents,* 2:1302–8.

Tewari, D. N. *Monograph on Eucalyptus.* Dehra Dun: Surya, 1992.

"Thai Government Studying Protesters' Complaints." Reuters, 22 April 1996.

"This Isn't Pulp Fiction." *Luxner News, South American Report* 2, no. 8 (1997): 25.

Thompson, Kenneth. "The Australian Fever Tree in California: Eucalypts and Malaria Prophylaxis." *Annals of the Association of American Geographers* 60 (1970): 230–44.

Troup, R. S. *Exotic Forest Trees of the British Empire.* Oxford: Clarendon Press, 1932.

Turnbull, J. W., and L. D. Pryor. "Choice of Species and Seed Sources." In *Eucalypts for Wood Production,* ed. Hillis and Brown, 6–65.

Tyrrell, Ian. *True Gardens of the Gods: Californian-Australian Environmental Reform, 1860–1930.* Berkeley: University of California Press, 1999.

Ummayya, Pandurang, and Bharat Dogra. "Planting Trees." *Ecologist* 13 (1983): 186.

Uniyal, Mahesh. "India-Environment: Industrial Forests Scheme in Trouble." Inter Press Service, 19 July 1995.

Urry, Maggie. "World Pulp and Paper Industry." *Financial Times,* 13 December 1989.

U.S. State Department. "Uses of the Eucalyptus Tree." Consular Reports 168. Washington, D.C.: Government Printing Office, 1894.

Vale, Antonio B. "Production Goals for Eucalyptus Plantations in Brazil." Ph.D. dissertation, University of Washington, Seattle, 1979.

"The Vegetation of the Island of St. Léger in Lago Maggiore." *Journal of the Royal Horticultural Society* 38 (1912–13): 503–14.

Vercoe, T. K. "Australian Trees on Tour." In *Australasian Forestry and the Global Environment,* ed. R. N. Thwaites and B. J. Schaumberg (Alexandra Headland: Institute of Foresters of Australia, 1993), 187–94.

Vilmorin, Henry de. "Flowers of the French Riviera." *Journal of the Royal Horticultural Society* 15 (1894): 80–104.

Von Mueller, Ferdinand. "Anniversary Address." *Proceedings of the Royal Society of Victoria* 4 (1859): 1–8.

———. *Eucalyptographia: A Descriptive Atlas of the Eucalypts of Australia.* 10 vols. Melbourne: Ferres, 1879–84.

———. "On the General Introduction of Useful Plants into Victoria." *Proceedings of the Royal Society of Victoria* 2 (1857): 93–109.

———. *Select Extra-Tropical Plants.* Melbourne: Ferres, 1885.

Waghorn, Jane. "The Debate Goes On . . . : Fifth Annual Non-chlorine Bleaching Conference." *World Paper* 220 (July 1995): 32–37.

Warren, Viola L. "The Eucalyptus Crusade." *Southern California Quarterly* 44 (1962): 31–42.

Weeks, Scott. "When Will It End?" *Latin Finance* 70 (1995): 49.

White, K. J. "Silviculture of *Eucalyptus* Plantings—Learning in the Region." In *Proceedings of the Regional Expert Consultation on Eucalyptus,* 3 vols. (Bangkok: FAO, 1995), 1:73–89.

"William Saunders." In *USDA Yearbook for 1900* (Washington, D.C.: Government Printing Office, 1901), 625.

Willis, M. *By Their Fruits: A Life of Ferdinand von Mueller.* Sydney: Angus and Robertson, 1949.

Wood, H. Trueman, ed. *Reports on the Colonial Section of the Exhibition, Colonial and Indian Exhibition.* London: Clowes, 1887.

References

Woodson, Weldon D. "Eucalyptus Boom and Bust." *Railroad Magazine* 52 (1950): 74–79.

World Bank. *The Forest Sector.* Washington, D.C.: World Bank, 1991.

"World Pulp Prices." *Asia Pulse* 12 (May 1998).

Wyman, Vic. "Spending Roller-Coaster." *Pulp and Paper* 37 (1995): 28.

Zacharin, Robert F. *Emigrant Eucalypts: Gum Trees and Exotics.* Melbourne: Melbourne University Press, 1978.

Zobel, Bruce J. [Address to Aracruz Researchers.] In *The New Eucalypt Forest: Proceedings of the Marcus Wallenberg Foundation Symposia* (Falun, 1984).

———. "Eucalyptus in the Forest Industry." *Tappi Journal* 71 (1988): 42–45.

Zobel, Bruce J., Gerrit Van Wyk, and Per Stahl. *Growing Exotic Forests.* New York: John Wiley and Sons, 1987.

Zon, R., and W. N. Sparhawk. *Forest Resources of the World.* Vol. 2. New York: McGraw Hill, 1923.

Index

ABECEL, 109
ACIAR, 163, 164
afforestation, 54, 135, 145, 146, 148, 178. *See also* reforestation; revegetation
Africa, 64, 107, 108, 140, 147, 168; eucalyptus density in, 103, 166; eucalyptus promotion in, 159; plantations in, 150, 152; sub-Saharan, 7; West, 57, 195n1b. *See also* South Africa
Agriculture, U.S. Department of, 54, 85, 159–60
Aiton, William, 26
Albuquerque, Frederico de, 97
Algeria, 32, 37, 39–40, 54
Anderson, William, 25, 26
Aracruz Celulose S.A., 108–18, 148, 150
Argentina, 58, 66, 81, 110; research in, 37, 99, 103; species in, 7, 62, 102, 116, 206n16
Arizona, 68, 73, 74
Asia, x, 108, 115, 147, 159
ATSC, 162, 163
Australia, x, xi, 1–3, 12–23, 77, 149, 198n9; biodiversity in, 151, 152–53; growth rates in, 173–74; markets in, 26–27, 162; oil production in, 8–9, 51, 78, 84; research in, 7, 98, 100, 106, 168, 204n26; South, 11, 34; tannin industry in, 196n18; Western, 18, 34, 163. *See also specific territories*
Australian Center for International Agricultural Research (ACIAR), 163, 164
Australian Development Bureau, 133, 162
Australian Tree Seed Centre (ATSC), 162, 163

Balfour, Arthur James, 33
Balfour, James Maitland, 33
Bandyopadhyay, J., 137, 138
Banks, Sir Joseph, 8, 10, 24, 25, 28, 38, 50
bastard mahogany, 53
Behr, Hans Hermann, 66

Bentley, R., 40–41
Betts, H. S., 83–84
Bhattacharya, Pranab K., 141
Birkbeck, Robert, 43
Birla Institute of Research, 154
blue gum. See *E. globulus*
blue mallee, 9
Boland, Douglas, 104, 162–63
Bonpland, Aimé, 45
Bosisto, Joseph, 8, 37, 51–53
botanical gardens: Assam (India), 129; Brazil, 97; Edinburgh (Scotland), 28; Kew (England), 26, 28, 30, 33, 35, 38; Melbourne (Australia), 45, 48, 50, 54, 60, 62; Naples (Italy), 46; Oxford (England), 28
Botanical Society of Edinburgh, 33, 39
boxes, 10
Brazil: environmental issues in, 116–18, 145, 177; eucalyptus density in, 103, 104, 118, 166; markets in, 27, 86, 173; plantations in, 148–49, 153, 178; promotion in, 66, 97, 160; pulp production in, 108, 110, 116, 117, 118, 120; research in, 100, 103–4, 106, 111; species in, 2, 62, 107, 108, 112; spread of eucalypts in, 68, 96–104, 109, 175. *See also* Aracruz Celulose S.A.; *and specific cities*
Brazil alba, 112. See also *E. urophylla*
Brazilian Environmental Agency, 116
Brazilian Pulp Exporters Association (ABECEL), 109
British Medical Journal, 56
Brunel, Adolph, 57, 199n34
Busch, John, 28

Calder, I. R., 155
California, 9, 10, 66–95; adaptability in, 42, 82, 85, 99; eucalypts as ornamentals in, 66–67, 174; landscape of, 88, 90–94, 160;

Index

Index

Index

Index

United States, 7, 98, 100, 102, 148; eucalyptus products in, 76, 106; markets in, 52, 113; spread of eucalypts in, xii, 41, 54, 66, 70, 159–60. *See also* California; *and specific organizations*
Uruguay, 66, 97, 99, 102, 116, 152
"Utilization of California Eucalyptus" (Betts and Smith), 83

van Wyk, G., 155
Victoria (Australia), 3, 4, 25, 34, 48, 50, 60, 82
Victoria (queen of England), 37, 51
Vienna, 52
Vietnam, 163–64
von Mueller, Ferdinand, 76, 101; on improving nature, 48, 70; on medicinal properties, 37, 39, 57–58; promotion of eucalyptus by, 45–46, 48–59, 60, 129, 172; research of, 33, 48–54, 70, 71, 72

Walker, William C., 67
White, John, 8
white peppermint, 46, 58, 199n38
Whittinghame hybrid, 33
wood production, 4, 58, 114–15, 165, 167, 173, 176; in Brazil, 97, 98; in California, 81–86;

and classification, 10, 11; demand for, ix, 67–68, 71, 131, 144, 145, 146, 147, 161; in developing nations, 159; and environmental issues, 117, 144, 157; and exports, 47; and growth rates, 39; and hardiness, 30, 43; in India, 129, 130, 132, 136, 137, 178; in Latin America, 102; and quality, 35–36, 53, 73, 76–77, 81–86, 160; research on, x–xi, 174; in Thailand, 170, 171; von Mueller on, 54
wood yields, xi, 112, 198n9; and climate, 115; and paper production, 27, 119; on plantations, 147–48, 149, 207n20; research on, 30, 163
World Bank, ix, 137, 139, 140, 176
World Rainforest Movement, 117–18
World Symposium on Man-Made Forests (1967), 167

yellow box, 9, 32
yellow gum, 10–11, 32

Zacharin, Robert Fyfe, 43
Zimbabwe, 152, 168
Zobel, B. J., 155

About the Author

Robin W. Doughty was born in Hull, England, in 1941. He completed a B.A. degree in philosophy, *magna cum laude,* at Vatican University in Rome and a B.A. degree in geography, with honors, at Reading University, and received an honorary M.A. degree from the University of Oxford. He earned his Ph.D. in geography at the University of California, Berkeley. He is the author or co-author of seven previous books, among them *The Return of the Whooping Crane* (Austin: University of Texas Press, 1989), *The Mockingbird* (Austin: University of Texas Press, 1988), and *Wildlife and Man in Texas: Environmental Change and Conservation* (College Station: Texas A & M University Press, 1983). He has also published many articles in leading professional journals. He is a professor of geography at the University of Texas, Austin.

Library of Congress Cataloging-in-Publication Data

Doughty, Robin W.
    The eucalyptus : a natural and commercial history of the gum tree /
Robin W. Doughty.
        p.   cm. — (Center books in natural history)
    Includes bibliographical references and index.
    ISBN 0-8018-6231-0 (alk. paper)
    1. Eucalyptus History.   I. Title.   II. Series.
    SD397.E8D68   2000
    634.9'73766— dc21   99-34202
                                                          CIP